Our Human Story

Our Human Story

LOUISE HUMPHREY AND CHRIS STRINGER

Published by the Natural History Museum, London

To Stephen and Caroline

First published by the Natural History Museum, Cromwell Road, London SW7 5BD
© The Trustees of the Natural History Museum, London, 2018
Reprinted with updates 2019 and 2022

All rights reserved. No part of this publication may be transmitted in any form or by any means without prior permission from the British publisher.

The Authors have asserted their rights to be identified as the Authors of this work under the Copyright, Designs and Patents Act 1988.

ISBN 978 0 565 09391 4

A catalogue record for this book is available from the British Library

10 9 8 7 6 5 4

Design by Mercer Design, London

Reproduction by Saxon Digital Services

Printing by Toppan Leefung Printing Limited

Contents

CHAPTER 1	Our closest relatives	7
CHAPTER 2	The first three million years	21
CHAPTER 3	*Australopithecus*	35
CHAPTER 4	*Paranthropus*	71
CHAPTER 5	The origins of humans (genus *Homo*)	83
CHAPTER 6	*Homo sapiens*	139
	Further information	155
	Index	155
	Picture credits	160
	Acknowledgements	160

CHAPTER 1

Our closest relatives

Charles Darwin recognized that the closest living relatives of modern humans are the African apes – gorillas and chimpanzees – and predicted that the earliest fossil relatives of humans would be found in Africa. Genetic comparisons have now shown that humans are most closely related to chimpanzees, a group comprising the common chimpanzee and the bonobo (the lineage leading to the gorilla followed a separate path). Our human story begins some 7 million years ago, at the point where the human evolutionary path split off from the lineage leading to chimpanzees. Since then species on both lineages have continued to evolve as they adapted to diverse and changing environments.

The focus of this book is the human lineage, represented by a diverse group of species known as hominins. Over the last 150 years, since the Neanderthals were recognized as a separate species from *Homo sapiens*, scientists have named more than 20 hominin species. At least half of these species are based on fossils unearthed in the last 30 years, and the pace of new discoveries shows no sign of slowing down in the twenty-first century.

OPPOSITE Skeletons of *Homo erectus* (left), *Australopithecus afarensis* (middle) and *Homo sapiens* (right). All these species are hominins, a term that refers to modern humans and all species that are more closely related to modern humans than to chimpanzees. All known hominin species have some anatomical features that suggest that they walked upright on two legs.

DNA AND GENOMES

Deoxyribonucleic acid (DNA) is the molecule that lies at the centre of the transmission of genetic data from parents to their children. It is made up of sets of codes, called genes, which lie behind bodily traits like eye colour and blood groups. Our DNA also consists of much larger amounts of non-coding portions, some of which may play an important regulatory role in the operation of the coding segments. DNA must be copied on from one generation to the next in the process of reproduction, and the copying process can introduce errors. If the changes are not lethal, these 'mutations' can in turn be copied on to succeeding generations, which means genetic lineages can be recognized and traced through time. For some parts of the genetic code that mutate at fairly predictable rates, it may also be possible to estimate the time taken to accumulate mutations, providing a kind of molecular clock. The first and largest component of DNA is contained in thread-like bodies called chromosomes in the nucleus of body cells, and is often called nuclear or autosomal DNA. This contains the blueprints for most of the constituents of our body structure, and we inherit a blend of autosomal DNA from our parents, each contributing about 50% to our genetic makeup. The special chromosomes that determine the sex of a human (X, and the male-specific Y) also contain their own smaller amounts of DNA. DNA from the Y-chromosome can be used to trace evolutionary lineages through the male line only, as it is transmitted only from fathers to their sons. Body cells also contain energy-generating organelles called mitochondria, which have their own distinct and separately inherited mitochondrial DNA (mtDNA) that is passed from a mother to her children, via her eggs. The total package of genetic material transmittable to the next generation is called a genome and it is estimated that the human genome contains about 20,000 actual coding genes.

Using molecular clocks, it is estimated, for example, that the human and chimpanzee lineages diverged between about 5–9 million years ago, modern humans shared a last common ancestor with Neanderthals and Denisovans between about 500–750,000 years ago, and *Homo sapiens* has mainly differentiated into its constituent populations within the last 150,000 years.

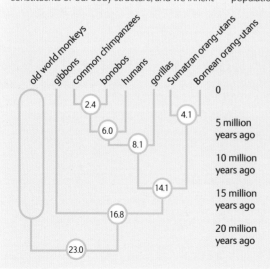

LEFT Using the rate of accumulation of changes in parts of the genetic code it is possible to produce molecular clocks, setting them by using known events in the fossil record, or through direct measurements of the mutation rate in recent organisms. Here is one example, based on mitochondrial DNA, which provides an approximate calibration for the divergences of humans and our closest primate relatives. Other parts of the genome provide different estimates to the one shown here.

How to identify a hominin

There are numerous anatomical features found throughout the skeleton and teeth that distinguish modern humans from chimpanzees. These anatomical differences have accumulated over time, so that the differences between the earliest hominins and apes that lived at the same time would have been much less pronounced than those between humans and chimpanzees today. When evaluating a potential hominin fossil, scientists look for at least one feature that is present in modern humans and not present in chimpanzees or in a hypothetical ape-like last common ancestor of modern humans and chimpanzees.

Gorilla and chimpanzee skeletons and teeth share more anatomical features with each other than either species shares with modern humans. These features include prominent canines in the upper and lower jaws, moderately sized chewing teeth arranged in parallel tooth rows, teeth with a thin enamel coating, a more projecting face and a smaller brain size, and a skeleton that was adapted for climbing and travelling on the ground using all four limbs. For a long time scientists assumed that the last common ancestor of humans and chimpanzees – just before the two lineages diverged – would exhibit many of the anatomical features shared by chimpanzees and gorillas. However recent fossil discoveries reveal that some of the earliest hominin contenders have some anatomical features that differ from those of chimpanzees and more recent hominins. This anatomical diversity can make it even more difficult to distinguish between the earliest hominins and other fossil ape species that lived at the same time but were not part of the hominin lineage.

Many of the anatomical features that distinguish modern humans and chimpanzees are related to differences in posture and locomotion. Humans are bipedal; we walk and sometimes run on two legs, and this presents a unique set of challenges in terms of balance, weight transfer and energy expenditure. In a biped, the skull is balanced directly above the vertebral column, and the entire weight of the upper body is transferred through the lower back, pelvis and legs to the feet. To help maintain balance during walking and running, the centre of gravity is shifted closer to the midline of the body and above the feet. Anatomical changes to facilitate bipedalism occur from head to foot (see p. 23).

One of the most visually striking differences between humans and apes can be seen when we open our mouths. All living apes have large projecting canines in the upper and lower jaw. There is a gap (or diastema) between the incisor and canine in the upper jaw to accommodate the large lower canine when the jaws are closed. The

ABOVE This male chimpanzee is displaying large projecting canines in the upper and lower jaw.

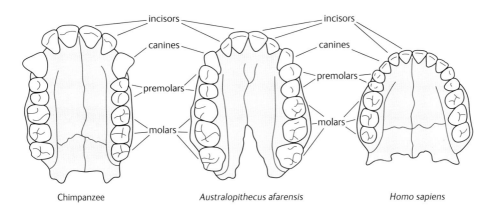

ABOVE The upper jaws of a chimpanzee, *Australopithecus afarensis* and a modern human reveal differences in canine size and the shape of the dental arcade. The premolars and molars are collectively known as cheek teeth.

upper canines wear against the front lower premolars, forming a sharpened edge on one side in a process known as honing. The anatomical features linked to honing are called the canine premolar honing complex. Humans have lost all traces of the canine premolar honing complex. We have small canines that function more like incisors and wear on their tips instead of on the sides. The gap between the canines and incisor in the upper jaw is normally closed as there is no need to make space for a projecting lower canine. Evidence for a reduction in canine size is one of the first signs that scientists look for when evaluating possible early hominin fossils.

Humans and apes have different tooth proportions and a differently shaped 'dental arcade'. The ape dentition is broader at the front due to the larger size of the front teeth (incisors and especially canines), and the cheek teeth behind the canines (premolars and molars) are arranged in two parallel rows, which results in a U-shaped dental arcade. In humans the canines and incisors are smaller and the teeth are arranged in a more rounded or parabolic dental arcade.

Another striking feature of modern humans is our large brains, which are surrounded by a smooth and rounded braincase. As brains do not fossilize, scientists measure the volume of the cranial vault (the bones that surround the brain) as a proxy for brain size. The hominin fossil record shows that a substantial increase in cranial capacity occurred relatively recently, in the last 2 million years, so this characteristic is less useful for identifying the earliest hominins.

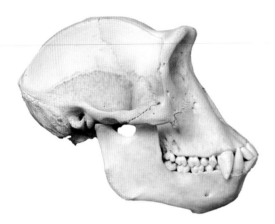

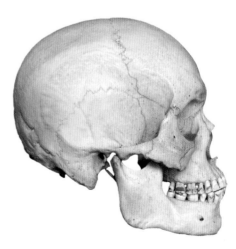

ABOVE The skulls of a chimpanzee, *Australopithecus africanus* and a modern human reveal differences in facial projection and the size and shape of the cranial vault.

Apes use their hands for a variety of tasks including grooming and food acquisition, for example, picking fruit or poking a stick into a termite mound. They also use their hands for locomotion, when moving around in the trees or on the ground. Apes have long curved fingers with strong muscles that are well suited for climbing, swinging or hanging from branches. When travelling on the ground, they use their hands as well as their feet. Chimpanzees and gorillas walk on the backs of their fingers (this is called knuckle walking) whereas orangutans fold their hands into fists and walk on the sides (called fist walking). Some fossil apes appear to have placed their weight on the palms of their hands when moving along branches, like modern monkeys. Because of the variation seen in living and fossil apes, scientists can't be sure that the last common ancestor of chimpanzees and humans was a knuckle walker.

LEFT Chimpanzees use a style of locomotion known as knuckle walking when travelling on the ground.

The human hand has been freed from the constraints of locomotion, and is better adapted for precise and powerful gripping and controlled manipulation of objects. The finger bones of the human hand are less curved and shorter than those of chimpanzees, so they are less well suited for hanging from branches. Conversely the thumb is strong and relatively long and better able to achieve a firm grip with each of the finger tips. Other clues for improved manipulative abilities in the human hand include flattened and widened tufts at the end of the finger bones to support our broad fingertips.

THE RARE PROCESS OF FOSSILIZATION

Very few animals become part of the fossil record because fossilization requires an unusual set of circumstances. Often animals are killed by predators or their carcasses are located by scavengers that set about devouring the body. Soft tissues are stripped away by teeth or beaks and softer parts of the skeleton may be crushed by powerful jaws. Some carnivores steal away bits of the carcasses and stash them out of reach in dens or in trees before they are eaten. Any remaining bits of the carcass are gradually dispersed over a wide area, and many of these scattered bones eventually disintegrate into unrecognizable fragments due to the harsh environment and pressures of passing hooves. However, occasionally the body, or surviving bits of it, might be quickly covered by sediment or dropped into an inaccessible place, remaining undisturbed for millions of years. This is when fossilization may occur. Many hominin fossils from open air sites, such as those in eastern and central Africa, have been found close to the edges of ancient lakes and rivers as these waterside locations provide an environment that is favourable for rapid burial and subsequent fossil formation. Under rare circumstances a whole body may sink into mud or be covered by a silty deposit, providing the potential for future recovery of a complete fossil skeleton (see p. 55 for example). Once buried, bones and teeth start to take up chemicals from the surrounding sediment. The chemicals replace the original material, transforming the remains into a fossil.

Many of the key early hominin sites in South Africa, such as Sterkfontein or Malapa, are caves or collapsed cave systems. Bones and other objects may have been washed or dropped into the cave system from above, or dragged in by predators such as hyenas, or bone-collecting animals such as porcupines. In a few exceptional cases relatively complete hominin skeletons have been found, and these may be the remains of individuals who accidently fell into the cave system through concealed chimneys or entered deliberately in search of water or shelter, and became trapped and died. Over time the bones and other debris that accumulate in underground chambers and crevices become cemented together into mineral conglomerates known as breccia. The fossils are sometimes exposed during mining activities or engineering work or found by cavers.

Some more recent hominin species, such as modern humans and Neanderthals, buried their dead or deposited bodies in places that were difficult for scavengers to access. This greatly increases the chance of preservation and the likelihood of fossilization and recovering a complete skeleton.

Recognizing hominin species

It is easier to identify a group of living animals that belong to the same species than to recognize species in the fossil record. The most widely used definition of a species is that it is 'a group of interbreeding natural populations that are reproductively isolated from other such groups'. Such a definition is useful for biologists, as it can easily be observed in living animals. But it is impossible to observe in extinct species. For the time being, scientists have opted for a more pragmatic approach: hominins belonging to the same species should resemble each other more closely than they resemble members of a different species. In practice this rule is not always easy to apply and scientists do not always agree about how to divide up the hominin fossil record into species.

One of the problems facing scientists is how much anatomical variation to expect or to tolerate within a species. Fossil species may span hundreds of thousands of years and during that time they may change in response to selective pressures, such as changing climate, or genetic drift, introducing additional variation. Differences in shape and size between males and females, known as sexual dimorphism, can also contribute to variation in a species. Some species exhibit high sexual dimorphism, so variation in the size and shape of bones and teeth and in the expression of muscle attachments throughout the skeleton is higher than in species in which males and females are more similar.

Another difficulty is caused by deficiencies in the fossil record. When comparing fossils, scientists look for shared 'derived' anatomical features (anatomical features that have changed from the ancestral, or primitive, condition) as these demonstrate that fossils are descended from the same ancestor and are more closely related to each other than to other hominins who lack those anatomical features. The circumstances for fossilization to occur are rare (see opposite) and consequently there are many gaps in the record simply because specimens were not preserved or have not yet been found. Some species are represented by small and fragmentary fossil assemblages and characterized by only a few diagnostic anatomical features whereas other species have comprehensive fossil records. Scientists cannot make useful comparisons between fossils and identify shared derived traits unless there is significant anatomical overlap in what has been preserved.

The fossil record reveals that there are many different ways of being a hominin, and we can learn a great deal about this without knowing exactly how many species existed and how they were related to one another. Some of these hominins overlapped with one another in space and time but in most cases we will never know whether they encountered each other, and whether those meetings were co-operative, hostile or simply indifferent.

Reconstructing hominin diets

The food that a hominin chose to eat and the way in which it obtained that food are key aspects of its survival strategy. Scientists have a variety of techniques for reconstructing the diets of early hominins. The chemical composition of our body tissues reflects our diet and, in this sense, you really are what you eat. Analysis of carbon isotopes has transformed our understanding of early hominin diets. Carbon isotopes from fossil, or modern, tooth enamel reflect the foods that an animal ate while its teeth were developing. Biologists divide plants into two main groups based on the way they photosynthesize, calling them C3 and C4 plants. Carbon isotope values can distinguish between C3 and C4 foods and so by examining the isotope composition of tooth enamel, scientists gain an understanding of the animal's diet. The C3 foods include fruits, leaves and other edible parts of trees, bushes and shrubs. These foods are found in forest or woodland habitats. The C4 foods include plants that grow in more open savannah habitats. They include seeds, leaves, roots and tubers from sedges and tropical grasses. The carbon isotope signature is carried up the food chain, so that animals that eat C3 or C4 plants will also have a C3 or C4 signal in their body tissues. Therefore, hominins feeding on those animals will take up the signal. Chimpanzees eat an almost exclusively C3 diet. Evidence for consumption of C4 foods by a fossil hominin would imply a diversification of diet and suggest that they were living in a different type of environment.

Another way of determining an animal's diet is by looking at the surface of the teeth. Microscopic traces on the surfaces of fossil teeth, which are caused by biting, crushing and grinding different types of food, are known as microwear. Scientists examine the microwear on fossil hominin teeth to detect whether they were eating foods that were hard or soft or tough or fibrous.

Dental plaque also has a story to tell. Phytoliths are tiny silica bodies produced by plants that can become trapped inside dental plaque when the plants are eaten or chewed. They can be extracted from fossilized dental plaque (calculus) adhering to the surface of a tooth, and identified to reveal the types of plants that an animal was chewing. This technique has been used to investigate the diet of recent and ancient fossil hominins.

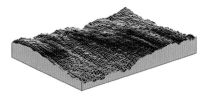 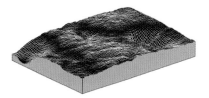

ABOVE Dental microwear texture analysis has revealed variation in early hominin diets. Both of the tooth surfaces shown are from *Homo erectus* and indicate a variable diet. The uniform scratches on the left are made by chewing tough foods such as meat or course vegetation, whereas the pits on the right suggest a diet that included hard objects such as nuts or bone.

 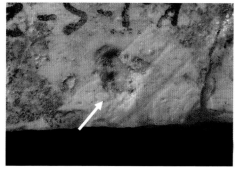

ABOVE Cut marks and percussion marks on these bones, such as these examples from Kanjera in Kenya, reveal that hominins were regularly processing animal carcasses to obtain meat by about 2 million years ago.

Cut marks and other damage caused by processing animal carcasses for meat and bone marrow are direct evidence for meat eating, although it may not always be clear which hominin species carried out these actions. However, the cut marks do tell us which animals were being eaten and may reveal whether the hominin that butchered the animal was a hunter or scavenger. If the cut marks were made after damage caused by carnivore teeth, it is more likely that the hominins were scavenging from a carcass of an animal that had been killed by a predator such as a lion or cheetah. Butchery marks found on hominin bones may suggest cannibalism, or the presence of another meat-eating hominin species. Tool types and wear on those tools also provide clues about how they were used to obtain and process food.

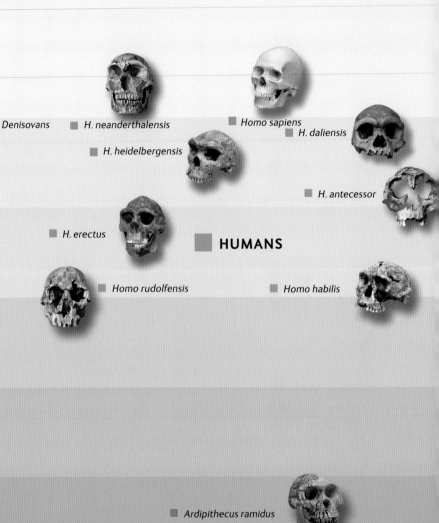

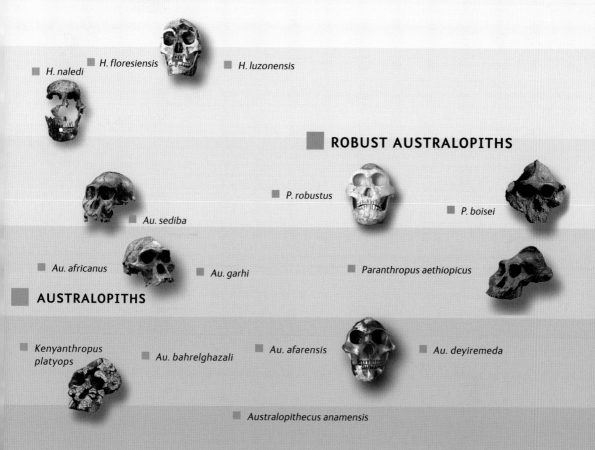

ABOVE The earliest known hominin fossils are nearly 7 million years old. Modern humans are now the only surviving hominin species, but until recently we coexisted with other hominin species.

CHAPTER 2

The first three million years

THE PERIOD BETWEEN 4.4 and 7 million years ago is represented by four species of hominin that have been placed in three different genera: *Sahelanthropus*, *Orrorin* and *Ardipithecus*. Fossils from each of these groups have at least one anatomical feature that could place them on the human lineage. Each of these early species is represented by different parts of the skeleton, with only a small amount of overlap, so it is not possible to make direct anatomical comparisons for many parts of the skeleton. With so few finds and no complete skeletons to compare, scientists are still debating whether there are indeed three genera, or whether all three groups should be lumped together in a single genus, *Ardipithecus*.

Sahelanthropus tchadensis

In 2002, Michel Brunet and his team announced the discovery of a new genus and species of early hominin, *Sahelanthropus tchadensis* (meaning 'Saharan hominin from Chad'). The fossil, which was found in central Africa, was remarkable both for its antiquity and its location. Anatomical features place *Sahelanthropus* close to the divergence of the human and chimpanzee lineages, and at approximately

OPPOSITE The Djurab desert in Chad has yielded fossils of *Sahelanthropus* dating from approximatively 7 million years ago, close to the estimated divergence date of the hominin and chimpanzee lineages.

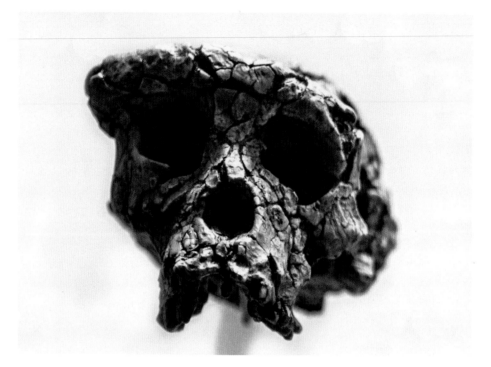

ABOVE The most complete specimen of *Sahelanthropus tchadensis* is a cranium known as Toumaï that was distorted during the process of fossilization.

7 million years old, the species also lived close in time to current estimates for the timing of this divergence.

Sahelanthropus fossils have been found at three localities in the Toros-Menalla in the Djurab desert, Chad. The fossils, which represent at least six individuals, include a remarkably complete but distorted cranium, fragments of a lower jaw and several loose teeth. The cranium was nicknamed Toumaï, meaning 'hope of life'. Before this discovery, the earliest known hominins had been found only at sites in eastern Africa. This new species therefore revealed that the earliest hominins were much more widely distributed across Africa than previously supposed.

The *Sahelanthropus* fossils were estimated to be 6–7 million years old using biochronology. This technique involves matching the types of animals found at an undated locality with similar assemblages from sites that have already been securely dated. The animal fossils found alongside *Sahelanthropus* were matched with those from well-dated sites from the Rift Valley in eastern Africa. Subsequently a more precise age estimation of 6.8–7.2 million years was obtained, based on the relative quantities of beryllium isotopes in sediment samples from the sites where the fossils were found. The dates confirm that *Sahelanthropus* is the earliest as well as the most primitive hominin contender.

The front teeth of *Sahelanthropus* provide the most convincing evidence of its hominin status. The canines are small and worn at the tip (apical wear facet), and there is no gap between the lower canine and premolar. Since bipedalism is considered to be one of the key defining features of a hominin, scientists carefully examined the base of the skull for evidence of an upright posture. They first had to produce a virtual reconstruction of the skull based on a scan of Toumaï, using a three-dimensional X-ray technique known as Computerized Tomography (CT). This allowed them to visualize the skull without the cracks and deformations that are present in the original fossil. The reconstruction suggested that the foramen magnum – the hole at the base of the skull through which the spinal cord runs – was further forward than in chimpanzees and was more horizontally orientated. These features are consistent with upright walking, implying that bipedalism arose soon after the divergence of the chimpanzee and human lineages.

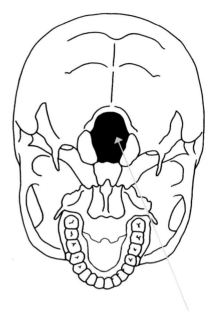

foramen magnum

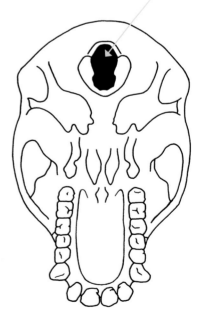

RIGHT The foramen magnum is the large hole in the base of the skull where the spinal cord enters. It is further forward in humans (top) than in apes (bottom) because of our upright posture.

Sahelanthropus displays a puzzling mix of anatomical features. One of the most striking characteristics of the face is a large and prominent brow ridge above the eye orbits. Large brow ridges are known in other hominins, but not until much later in the hominin fossil record, and their function is unclear. Apart from the brow ridges, Toumaï has an unexpectedly flat face and a small and low cranial vault, with an estimated brain size of about 360 cm^3, similar to that of a small chimpanzee.

No new *Sahelanthropus* fossils have been collected since the original discoveries in 2001 and 2002, but scientists have continued to study the fossil assemblage in the laboratory. They have identified a partial femur (thigh bone) and two arm bones that were found at the same locality as Toumaï and are thought to belong to the same species. Research on the shape and anatomical features of these bones is ongoing. Scientists are aiming to diagnose whether the femur displays any unambiguous features associated with bipedal locomotion.

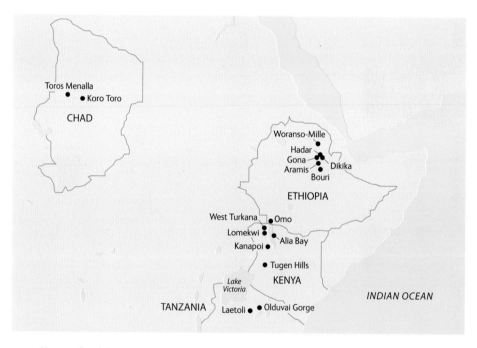

ABOVE Hominin fossils have been found at sites in Chad in central Africa and at numerous sites along the Rift Valley in eastern Africa. Some of the key sites for the earliest hominins and australopiths are shown here.

Fossils from a diverse range of mammals, reptiles and fish were found at the same site as Toumaï, including monkeys, giraffes, elephants, antelopes, crocodiles, hippos and otters. These indicate that *Sahelanthropus* was living in a river or lakeside environment surrounded by a mosaic of habitats including forest growing along the banks of a river, wooded savannah and grassland.

Orrorin tugenensis

Orrorin tugenensis is almost as ancient as *Sahelanthropus*. *Orrorin* is represented by a collection of fossils dated at around 5.7–6.2 million years old derived from four localities in the Lukeino formation in the Tugen Hills region of Kenya. The new genus and species was established by Brigitte Senut and colleagues in 2001. *Orrorin* means 'original man' in the Tugen language and *tugenensis* honours the region where the fossils were discovered.

ABOVE *Orrorin tugenensis* was named after the Tugen Hills region of Kenya, where the fossils were discovered.

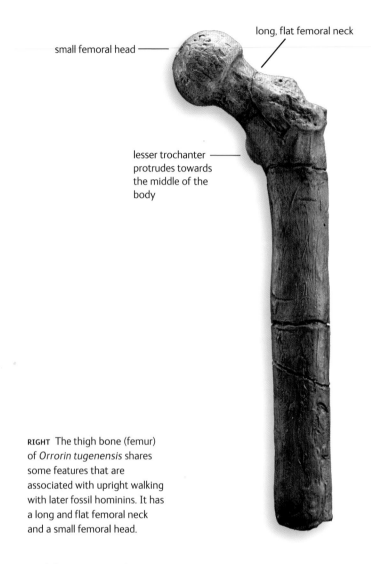

RIGHT The thigh bone (femur) of *Orrorin tugenensis* shares some features that are associated with upright walking with later fossil hominins. It has a long and flat femoral neck and a small femoral head.

small femoral head

long, flat femoral neck

lesser trochanter protrudes towards the middle of the body

The *Orrorin* fossil assemblage provides clues about the hominin's lifestyle and anatomy, but there are still uncertainties about its status and relationship to other species. The hominin status of *Orrorin* is supported primarily by the shape and anatomical features of the leg bones. The femur (thigh bone) has several traits that indicate bipedal walking, such as a long and flat femoral neck. It has a unique shape that shows similarities with both *Australopithecus* (see p. 35) and fossil apes from the Miocene.

The evidence from the dentition is more ambiguous. The upper canine of *Orrorin* is smaller than those of many apes but the shape of the crown resembles a female chimpanzee, and the anterior lower premolar, which forms part of the canine premolar honing complex, is also ape-like. The humerus (upper arm bone) has strong muscle attachments suggesting powerful arms that could be used for climbing and foraging in the trees. *Orrorin* is thought to have weighed about 45 kg (99 lb), but as there are no bones from the cranial vault, its brain size is unknown. *Orrorin* lived in a woodland habitat close to a lake or stream, and may have been foraging both in the tree canopy and on the ground.

Ardipithecus ramidus

The most complete assemblage of fossils from this period belongs to the genus *Ardipithecus*. The name was established by palaeoanthropologist Tim White and colleagues in 1995 and means 'ground living ape' in the Afar language. *Ardipithecus ramidus* (meaning 'ground living root hominin') is represented by an extensive collection of fossils from Aramis in the Middle Awash region of Ethiopia. Scientists launched a systematic palaeontological investigation of this region in 1992 and the first hominin fossils were discovered at Aramis vertebrate palaeontology locality 1 (ARA-VP-1) in December of that year. The fossil hominin assemblage from Aramis now includes parts of more than 30 individuals with one particularly impressive skeleton of an adult female, nicknamed Ardi. Palaeoanthropologists recovered more than 100 fossil fragments belonging to Ardi, including an almost complete set of teeth and parts of the skull and postcranial skeleton.

Cranial bones belonging to Ardi were found scattered over a wide area and in poor condition. A virtual model of the cranium was reconstructed from CT scans, revealing a small face and a cranial capacity comparable to small chimpanzee or bonobo. The cranial base was short, a feature shared with *Sahelanthropus*. Teeth from multiple individuals revealed that the upper canines of *Ar. ramidus* were short with tips that wore down to the level of the surrounding teeth. There was no evidence of a canine premolar honing complex.

The reconstructed skeleton of Ardi has revealed a unique and unexpected suite of anatomical features. The upper part of the pelvis was shorter and broader than those of living apes and would have prevented Ardi from lurching

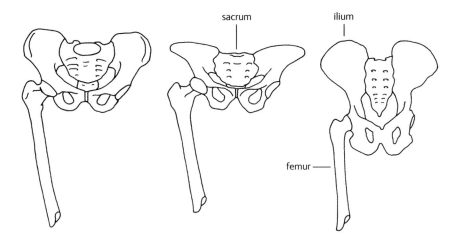

ABOVE The chimpanzee pelvis (right) has a long ilium positioned at the back. Bipedal hominins such as *Australopithecus* (middle) and modern humans (left) have a broad pelvis with a shorter ilium positioned on the side and a wider sacrum. The femur (thigh bone) of bipedal hominins is angled inwards towards the knee so that the centre of gravity lies directly above the foot.

from side to side during bipedal walking. However, the shape of the lower part of the pelvis resembled that of an ape and provided an anchor for the powerful leg muscles used during climbing.

The hands and feet present a unique mosaic of features, resembling neither chimpanzees nor humans nor an anatomically intermediate state. Ardi has long curving fingers that would have allowed her to grasp or hang from branches while climbing in the trees but the bones in her palm are shorter than those of a chimpanzee and her wrist was more flexible. Ardi would not have been able to manipulate objects with the precision needed to manufacture stone tools. One of the most unexpected features of Ardi's anatomy is in her feet. She has a widely divergent and opposable big toe, ideal for grasping tree branches but less well adapted for walking upright. The discovery of another partial skeleton of

OPPOSITE The oldest hominin skeleton, known as Ardi, has been reconstructed from more than 100 fossil fragments. She stood about 1.3 m (4¼ ft) tall, weighed about 50 kg (110 lb) and had a brain size comparable to a small chimpanzee. Scientists took several years to extract the fossils from the surrounding matrix using a needle under a microscope.

LEFT Artists reconstruction of a small group of *Ardipithecus ramidus* foraging in a forest clearing. Carbon isotope analysis reveals that this species ate a mainly C3 diet, that could have included fruits, leaves and other forest foods.

Ar. ramidus from the Gona study area of Ethiopia dated to between 4.6 and 4.3 million years old was announced in 2018. The Gona fossil possessed a foot that was better adapted for bipedal locomotion than Ardi, revealing an intriguing diversity within this species.

Ardipithecus ramidus shows a small degree of sexual dimorphism. Males have a slightly larger body size than females, similar to differences observed between male and female humans or chimpanzees and much less than the differences between male and female gorillas or orang-utans. Male canines are also only slightly larger than those of females. Low sexual dimorphism can indicate a social organization in which males are not in direct competition with each other, and this has been suggested for *Ar. ramidus*.

The fossil-bearing deposits at Aramis have been dated to approximately 4.4 million years ago, using radiometric dating techniques (see pp. 32-33). Thousands of animals and plant remains were collected from these deposits, allowing the research team to build up a detailed picture of the environment at this time. The species found alongside *Ar. ramidus* include leaf-eating monkeys and other forest animals and birds such as parrots and peafowl that prefer a wooded or forested habitat. The deposits also yielded seeds from fig and palm trees. Carbon isotopes reveal that *Ar. ramidus* ate a diet comprised mainly of forest foods such as fruits and leaves, with a small component of grasses or sedges.

Ardipithecus kadabba

A second, slightly earlier assemblage of *Ardipithecus* fossils was named as a separate species, *Ardipithecus kadabba* (meaning 'ground-living family ancestor' in the Afar language), by palaeoanthropologist Yohannes Haile-Selassie and colleagues in 2004. *Ar. kadabba* is represented by part of a lower jaw, teeth and postcranial bones from six localities in the Middle Awash study area of Ethiopia.

Most of the fossils that are reliably attributed to *Ar. kadabba* are between 5.4 and 5.8 million years old. A single tooth from Gona dated to 6.3 million years old has been tentatively assigned to this species. The argument for bipedalism in *Ar. kadabba* hinges on a single toe bone from the Sagantole Formation dated to 5.2 million years old, which has an upward angled joint surface. This shape resembles later bipedal hominins, but it has been recorded in fossil apes from the Miocene that were not bipedal, so additional evidence is needed to conclude that *Ar. kadabba* was bipedal. The dentition of *Ar. kadabba* is more primitive than *Ar. ramidus* and suggests that the canine premolar honing complex was still present, but possibly slightly modified. Other fossilized animals found at the same localities reveal that *Ar. ramidus* lived in a wooded habitat close to water sources such as streams or lakes.

Scientists are not yet certain where *Ardipithecus* fits into the hominin family tree. Some scientists consider *Ardipithecus* to be a plausible ancestor for *Australopithecus anamensis*, the earliest known species of *Australopithecus*. Others regard it as an extinct side branch. Some scientists have disputed whether *Ardipithecus* is a hominin or whether it is more closely related to another extinct group of apes. If this were the case, some of the anatomical traits shared between *Ardipithecus* and hominins would have evolved independently on both lineages.

DATING FOSSILS

In the simplest terms, there are two main ways to date fossil remains: relative dating and physical dating (also called radiometric, or less accurately, absolute dating). As the name implies, the first method only dates an object or layer in relation to another object or layer in a sequence of deposits. For example, the Toba volcano on Sumatra is known to have erupted about 74,000 years ago, spreading its characteristic ash for thousands of kilometres. A fossil that lies under that ash in a cave or lake deposit should, therefore, be older than the age of the eruption, whereas if it lies above the ash, it should be younger. Specimens lying within the ash layer in two different localities can be correlated – that is they should be of equivalent age, about 74,000 years. To move beyond relative dating, it's necessary to use a natural clock that can determine how much time has elapsed since an animal or plant died. Many of these natural clocks use physical properties like natural radioactivity to measure time. The most famous of such methods is radiocarbon dating, based around an unstable form of carbon called carbon-14 that decays with a half-life of about 5,730 years (that is, the amount present naturally halves by radioactive decay every 5,730 years). Cosmic radiation constantly creates this unstable isotope from nitrogen-13 in the Earth's upper atmosphere, and it is then incorporated into the bodies of living things, along with the more common and stable carbon-12. However, when the plant or animal dies, it stops accumulating radiocarbon, and the amount left begins to break down by radioactive decay. Thus, measuring the amount of carbon-14 left in, say, a fossil human bone can provide an estimate of the time since that human died.

BELOW A laboratory for radiocarbon dating in Oxford, UK.

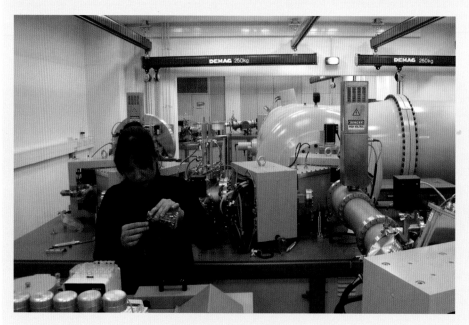

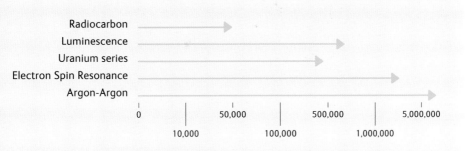

RANGES OF THE MAIN DATING METHODS FOR RECENT HUMAN EVOLUTION

Because the amount of measurable carbon-14 is increasingly small in material older than 30,000 years (since it is constantly breaking down), the method becomes less and less reliable, so it is fortunate that other radiometric methods have been developed to date both fossil and archaeological materials beyond the effective limits of radiocarbon dating. These include potassium–argon and argon–argon dating, applicable to volcanic rocks; uranium-series dating, especially effective on cave flowstones and corals; luminescence dating, used on sediments and heated stone tools; and electron spin resonance (ESR), applied to tooth enamel.

The first geologists relied entirely on relative dating to study sequences of rocks, as there were no reliable ways of determining their real ages. Now many methods can be used, including those described above. There is another technique that acts as both a relative and a physical dating procedure: palaeomagnetism. This makes use of the remarkable fact that the Earth's magnetic field periodically changes so that the North and South magnetic poles switch over. Fine-grained rocks such as clays and lavas containing iron can record the position of the Earth's poles at the time they were laid down. The last significant

ABOVE This diagram shows the approximate effective ranges of five physical dating methods.

reversal of the magnetic poles happened about 780,000 years ago, and there have been hundreds more such reversals in the Earth's history. Some of these reversals can be directly dated if they are recorded within volcanic rocks that are datable by potassium–argon or argon–argon methods. Such precise dating has been invaluable in building up detailed sequences in the geological timescale, a system of chronological dating that relates the Earth's geological strata to time. The Earth's record of its rocks is grouped into a number of Eras, with further subdivisions, called Periods or Epochs. Human evolution has taken place in the periods known as the Tertiary (because it was originally the third stage) and Quaternary (originally the fourth stage and includes recent and present time). The Tertiary Period is further divided, as is the Quaternary into the Pleistocene and Holocene (Recent) Epochs. In the last few years, some scientists have argued for the naming of a new stage after the Holocene, to reflect the huge impact that humans are now making on the planet – this epoch would be called the Anthropocene.

CHAPTER 3

Australopithecus

AUSTRALOPITHECUS IS A GENUS of small-bodied and small-brained bipedal early hominin species that are first known from around 4.2 million years ago and that still existed as recently as 2 million years ago. *Australopithecus* species were capable of upright walking but were not well adapted for travelling long distances. They had hands that were well suited for climbing trees and for the controlled manipulation of objects such as tools. Most species had larger molars than chimpanzees and are sometimes referred to as megadont ('big toothed'). It is likely that species in this group gave rise to two more recent hominin groups, *Homo* and *Paranthropus*, before 2.5 million years ago. They also overlap in time with the genus *Kenyanthropus* (see p. 62).

SOUTH AFRICAN *AUSTRALOPITHECUS*

Many of the key fossil sites in South Africa are from the Cradle of Humankind World Heritage Site, a landscape dotted with collapsed cave systems and potentially fossil-rich sites, less than an hour's drive from Johannesburg. The cave systems were formed millions of years ago by the action of groundwater and rainwater on the limestone in the dolomitic bedrock. When groundwater is acidic, it dissolves the limestone, forming deep, hollow chambers. Rainwater,

OPPOSITE *Australopithecus afarensis* is represented by a large number of fossils including these upper and lower jaws from Ethiopia.

which is also slightly acidic, percolates through cracks in the rock, dissolving the surrounding limestone to create larger cavities or expose the underlying chambers. Over time this produces a maze of underground caverns, sometimes connected to the surface by narrow vertical shafts. Underground caves and fissures may be open to the surface for a period of time and then close and reopen thousands of years later. Soil, bone and other materials accumulate in the caves and fissures when they are open and eventually become cemented together as breccia. Over time the roofs of the underground chambers can collapse forming sinkholes.

During the late nineteenth and early twentieth century the cave systems were heavily exploited by miners who extracted the limestone and converted it into lime for use in construction and gold processing. Many of the cave systems were blasted with dynamite, throwing up blocks of limestone and breccia and removing them from their geological context. At this time the miners and geologists who worked at the mines began to find the fossils of extinct animals. Some of these were collected and eventually found their way into the hands of amateur fossil hunters and scientists.

LEFT Key hominin sites in South Africa. Many fossil-bearing sites are clustered in the Cradle of Humankind, near Johannesburg.

ABOVE Sterkfontein has yielded many hominin fossils including the first complete *Australopithecus* skull, known as Mrs Ples, and the skeleton known as Little Foot.

The South African cave systems present a unique set of challenges at every stage of investigation, from the extraction of fossils to understanding how the assemblages accumulated and estimating their age. One of the methods used to date deposits at caves such as Sterkfontein and Malapa is palaeomagnetism. This method relies on the timing and sequence of changes in the Earth's magnetic field. Magnetically charged particles align themselves to the magnetic pole and retain this orientation when they are consolidated into rock. Therefore, the age of fossils can be estimated from the alignment of particles in the surrounding deposit. Another method is biochronology, which involves comparing the types of animals in a fossil assemblage with well-dated fossil assemblages from other sites such as those in eastern Africa. Scientists are also developing a range of radiometric dating methods that can be used to date breccia.

We know the South African fossil record includes hominin species belonging to the genera *Australopithecus*, *Paranthropus* and *Homo* and it is often assumed that each of these groups originated and diversified somewhere in eastern Africa, and that one or more species spread south, eventually reaching South Africa. It is possible that a more complete fossil record and better dating will one day reveal that some of these groups emerged in South Africa, or in another part of Africa altogether.

The earliest hominin fossils currently known from South Africa are from the Sterkfontein cave system and come from the Silberberg Grotto and Jakovec Cavern. These fossils may be equivalent in age to *Australopithecus anamensis* in eastern Africa, but are considered by some scientists to be more recent. The fossils from the Jakovec Cavern have not been assigned to a species because scientists do not have enough information to determine which of the existing hominin species they belong to, or whether they should be assigned to a new species.

Little Foot

In 1994, palaeoanthropologist Ron Clarke was looking through a box of miscellaneous fossils in the Sterkfontein field laboratory, when he identified four small hominin bones that had been overlooked previously. This find would lead to a remarkable chain of events, culminating in the discovery of an almost complete hominin skeleton. Clarke recognized the four bones as part of a hominin foot. The box of fossils that they came from had been extracted from a dump of rubble and breccia blocks close to the entrance of the Silberberg Grotto, Sterkfontein. Clarke published a description of the partial foot the following year and it was promptly dubbed Little Foot due to its diminutive size. The four bones suggested a foot that is human-like at the heel but becomes increasingly ape-like towards the toes. The shape of the big toe indicated that it would have diverged slightly from the other missing toes. Overall the shape of the bones was consistent with a bipedal foot that was capable of grasping.

Three years later, Clarke investigated the contents of another box labelled 'D18 Cercopithecoids', expecting to find monkey fossils from Dump 18 in the Silberberg Grotto. Among the fragmentary fossils he recognised further foot bones and the broken lower end of a leg bone belonging to Little Foot. One of the bones had a clean break and Clarke suspected that it might have been damaged during blasting by miners to open up the cave 60 years previously. Clarke gave a cast of the distal fragment of the right tibia to two of the Sterkfontein fossil preparators, Stephen Motsumi and Nkwane Molefe, and asked them to search the walls of the Silberberg Grotto for matching parts of the broken leg bone. Remarkably they found the matching pieces after two days of searching the walls with the aid of hand-held lamps.

Clarke and his team set about exposing the bones by chiselling away at the surrounding breccia. Painstaking excavation over the following weeks, months and years revealed most of an adult skeleton, including the skull, arms and hands, the right scapula and clavicle, several ribs and vertebrae, the entire pelvis, both legs and some more of the foot bones. The skeleton was lying on its back with its head tilted onto its right side and its left arm stretched out above its head, with the fingers folded over the upward facing palm of the hand. The left leg was crossed over the right leg. The position implies that the body rolled downslope before decomposition and that some slumping then occurred so that parts of the skeleton are at different levels. The completeness of the skeleton and the absence of carnivore damage on any of the bones suggest that the individual may have fallen into the cave system.

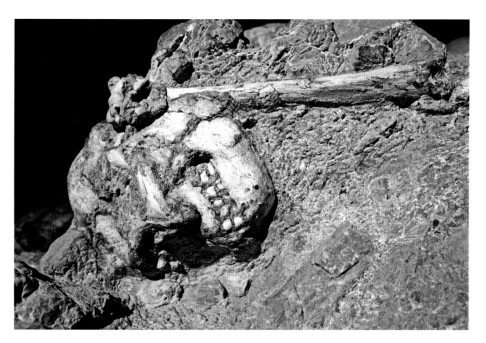

ABOVE The skull and upper arm bone of Little Foot were embedded in breccia in the Silberberg Grotto at Sterkfontein.

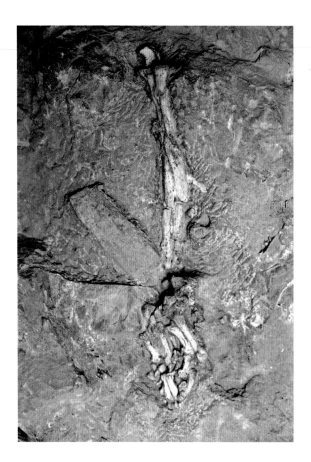

LEFT The left lower arm and hand of Little Foot were found alongside other parts of the skeleton.

Complete skeletons such as Little Foot present a rare opportunity to examine the cranial, postcranial and dental anatomy of a single individual. This provides scientists with conclusive evidence that a specific set of anatomical features occur in a single species. Little Foot is an early australopith and a series of studies have revealed details of its anatomy. The cranium has robust zygomatic arches (cheek bones) and a small sagittal crest (bony ridge) towards the back of the skull, both of which indicate strong chewing muscles. The skull had a small brain case, with an estimated brain size of 408 cm^3, which is at the lower end of the range of sizes observed in australopiths. The small canines and shape of the pelvis reveal that Little Foot was female and other features suggest an older adult. The leg bones were adapted for habitual upright walking whereas the arm and shoulder bones had traits associated with arboreal behaviors. The bones of the shoulder girdle were arranged in a way that is well suited to movements that require positioning the lower arm above the head such as climbing and suspension.

Dating has been controversial because the skeleton is thought to be older than the rock that surrounded it. Radioisotopic dating of sediments associated with the skeleton yielded an age estimate of 3.67 million years. This date implies that Little Foot lived at the same time as *Australopithecus afarensis* in eastern Africa, demonstrating that *Australopithecus* was already present in both eastern and southern Africa at this time. Clarke considers Little Foot to belong to *Australopithecus prometheus*, a name first given to a cranial bone discovered in the Makapansgat lime works in South Africa by Raymond Dart in 1948. However, not all researchers recognize *Au. prometheus* as a valid species.

Australopithecus africanus

Raymond Dart arrived in South Africa in 1923 having been persuaded by his mentors at University College London to take the position of Professor of Anatomy at the recently established University of Witwatersrand. The following year Dart was presented with a fossil baboon skull that had been found by miners working at the Buxton lime works near the village of Taung. Dart asked a colleague to send him any other interesting fossils recovered during mining operations and towards the end of 1924, he was rewarded with two crates of fossils. According to his own account, he spotted an endocranial cast at the top of the rock heap as soon as he opened the box and 'knew at glance that what lay in my hands was no ordinary anthropoidal brain'. The natural endocast – a fossilized mould of the interior of the cranial vault – fitted neatly into a limestone block that contained part of a fossilized skull. Dart started the painstaking process of extracting the fossilized bones from the surrounding breccia, and some weeks later the face and mandible of a small child were revealed. The child's jaws contained a complete set of milk teeth, with the first permanent molars emerging at the back of the mouth.

Dart announced the fossil in 1925 and gave it the name *Australopithecus africanus*, which means 'the southern African ape'. Dart argued that features of the preserved endocast suggested a more upright posture and bipedal walking. He recorded that the canines were small, projecting only slightly beyond the other teeth and that there was no gap between lower premolar and canine. Dart's publication was poorly received, with critics dismissing the find as an extinct great ape. The small brained putative hominin from Africa ran counter to the

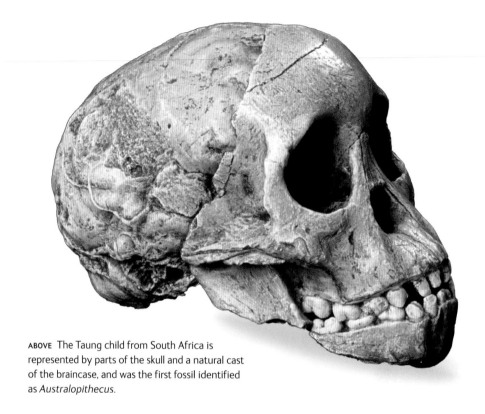

ABOVE The Taung child from South Africa is represented by parts of the skull and a natural cast of the braincase, and was the first fossil identified as *Australopithecus*.

prevailing view in the 1920s, which asserted that humans had originated in Europe or Asia and early hominins would be recognized by their large brains. This view persisted for more than 20 years but was eventually overturned as other fossil evidence accumulated.

More recent research on the Taung child has revealed intriguing details about how it lived and died. Scientists have carefully inspected the teeth under a microscope and discovered an area of defective enamel on each of the first permanent molars. This type of defect occurs when the normal process of enamel formation is temporarily interrupted. It suggests that the Taung child experienced a period of severe illness or trauma during early childhood when the tooth enamel was forming.

Scientists have also deciphered the likely cause of death. Scratch marks and breakage on the skull and eye sockets of the Taung child resemble the damage caused by bird talons. No other hominin fossils have ever been found at Taung,

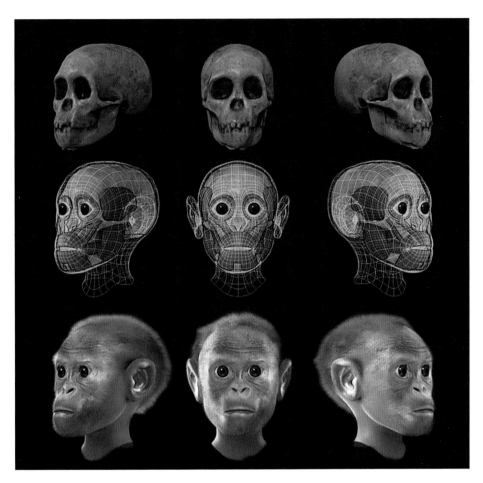

ABOVE This digital reconstruction of the Taung child was produced by building muscles (shown in red) onto the bone structure using forensic techniques.

but the site contains an unusual assemblage of fossils from small-bodied animals such as baboons and antelopes, hyraxes, tortoises and crabs. Some of these fossils bear similar marks. A likely scenario is that the Taung child was snatched away by a large predatory bird such as an eagle, which carried the small hominin to its nest with the rest of its prey.

Australopithecus africanus is now known from hundreds of fossils from other South African sites including Sterkfontein, Makapansgat and Gladysvale.

Fossil-bearing breccia from these sites are dated between approximately 2.4 and 3 million years old, but *Au. africanus* may well have been present in earlier and later periods. Much of what we know about the anatomy of *Au. africanus* was already known in the 1940s, owing to the discovery of an impressive collection of fossils from Sterkfontein by Robert Broom, a palaeontologist working at the Transvaal Museum in Pretoria. Among these were several almost complete crania and a partial skeleton that included the pelvis, much of the vertebral column, some ribs and the upper part of a femur (thigh bone). These discoveries persuaded most of the scientific community that bipedal walking and a reduction in canine size arose before any significant increase in brain size during hominin evolution.

The cranium and dentition of *Au. africanus* display a mix of derived and more primitive features. *Au. africanus* has small incisor-shaped canines and they are worn on the tips rather than on the sides. There is no gap between the upper permanent canines and incisors as there is no need to make space for a large lower canines. The front lower premolars each have two main cusps and were used for crushing and grinding food rather than sharpening the upper canines. All of these features imply a loss of the canine premolar honing complex. The chewing teeth are slightly larger than those of *Au. afarensis*, an earlier *Australopithecus* species from eastern Africa (see p. 52).

Australopithecus africanus has a low cranial vault with a receding forehead, as would be expected in a small-brained hominin. The average cranial capacity is 460 cm^3, suggesting that *Au. africanus* had slightly larger brains than chimpanzees or earlier *Australopithecus* species such as *Au. afarensis*. Male gorillas have larger brains than this, with cranial capacities averaging 535 cm^3, but they have a much larger body size, probably weighing nearly four times as much as *Australopithecus* adults.

The foramen magnum – the hole at the base of the skull through which the spinal cord runs – is shifted forward and orientated more horizontally than in apes, indicating that the head was balanced above the vertebral column as in other upright walking hominins (see p.23). The partial skeleton and other bones found at Sterkfontein reveal additional features associated with bipedal walking. The iliac blades of the pelvis are short and flare outward, which makes it easier to balance on one leg at a time during bipedal walking (see p.33). The wedging of the vertebrae reveals that the lower back was curved inwards at the base (this is known as lumbar lordosis), which helps to absorb shocks during bipedal walking. The consensus now is that *Au. africanus* could walk bipedally, but was still well adapted for climbing trees.

Australopithecus sediba

Malapa is a collapsed cave system not far from Sterkfontein in South Africa. The cave was blasted by miners in search of lime in the early twentieth century but they quickly abandoned Malapa in favour of more productive locations, leaving blocks of cave breccia strewn across the surrounding landscape. In 2008, Matthew Berger, then aged 9 years old, discovered a fossil clavicle embedded in one of the blocks of rubble that the lime miners had left behind. Since then Malapa has yielded fossils from at least four hominins, including partial skeletons belonging to a juvenile male and an adult female. This was the first time that two relatively complete partial skeletons of an early hominin species have been found at a single site. The fossils from Malapa have been precisely dated to just less than 2 million years old based on uranium-lead dating and palaeomagnetism.

Malapa means 'homestead' but the cave must once have been a death trap for unwary hominins and other animals that could have fallen into the underground chamber through concealed vertical shafts. Fossils representing at least 25 species

BELOW Two remarkably complete skeletons of *Australopithecus sediba* were discovered at Malapa in South Africa in 2008.

ABOVE The teeth and skeleton of the young male *Australopithecus sediba* known as MH1 were not fully developed when he died.

have been found at Malapa, including the brown hyena, wild dog, sabre-toothed cat, antelope and horse. As none of the fossils show signs of carnivore damage, the bodies of the animals that died in the underground chamber must have been out of reach of scavengers. Both hominin skeletons show perimortem bone injuries (injuries that that occurred at around the time of death) that could have been caused by falling into the cave or being hit by a rock.

In 2010, palaeoanthropologist Lee Berger (Matthew's father) and colleagues established a new species, *Australopithecus sediba* (meaning 'wellspring' in Sotho), for the Malapa hominin assemblage. *Australopithecus sediba* is defined on the basis of a unique combination of anatomical features that have not been found together in any other hominin species. It was a small-bodied and small brained biped similar in many respects to other *Australopithecus* species but also shared features with more recent hominins that have been placed in the genus *Homo*. The overall pattern that emerges from studies of different parts of the skeleton is that the upper part of the body and arms retained more primitive characteristics whereas the lower part of the body and the hands show more derived features. The juvenile male has a minimum cranial capacity of 420 cm^3, which is well within the range of other *Australopithecus* species.

The bones from the legs, feet and ankles of *Au. sediba* show a mosaic of primitive and derived features that indicate tree-climbing abilities and a unique form of bipedal walking. Reconstruction of the vertebral column revealed extreme lordosis (an inward curve of the lower part of the spine), a clear adaption to bipedalism, whereas other features of the spine indicated powerful trunk muscles associated with climbing. The arm and hand were well suited for climbing. The hand has curved fingers and muscle attachments that indicate a powerful grasp. The fingers are relatively short compared to the very long thumb, which enables thumb-to-finger precision gripping and controlled manipulation of small objects. Although there is no evidence so far for tool making at Malapa, stone tools occur at other nearby sites, and *Au. sediba* may also have been making tools from materials such as wood that do not normally survive in the archaeological record.

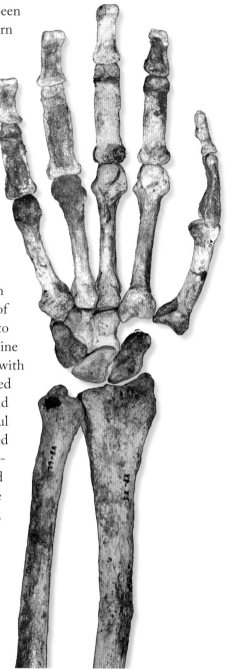

RIGHT The hand of *Australopithecus sediba* is almost complete and the bones reveal a hand capable of precision gripping and careful manipulation of objects.

Analysis of the chemical composition of tooth enamel from *Au. sediba* reveal a diet based on fruits, leaves or other edible parts of plants that grow in a woodland or forest environment. This was unexpected as all other hominins from this period and most *Australopithecus* species are known to have eaten a mixture of foods from forest and more open savannah habitats. Scientists were able to identify some of the plants that *Au. sediba* had been chewing from phytoliths trapped inside calculus deposits on their teeth. The phytoliths revealed a range of different plant foods including leaves, fruits and more surprisingly tree bark.

EASTERN AND CENTRAL AFRICAN *AUSTRALOPITHECUS*

Most of the fossil-bearing sites in eastern and central Africa are open air localities that have been exposed by erosion. Some fossils are found in layers of sediment

BELOW An exposed tuff at Oldupai Gorge in Tanzania. Tuffs are formed by consolidation of volcanic ash.

sandwiched between layers of compacted volcanic ash called tuffs. Each tuff has a unique chemical profile, which allows researchers to trace the same tuff for hundreds of kilometres across a landscape. The age of volcanic deposits can be accurately determined using radiometric techniques. The date of the underlying tuff represents the maximum age and the date of overlying tuff represents the minimum age for a fossil sandwiched between the two. Finding hominin fossils in this type of environment involves scouring the landscape in search of fossils that are starting to erode out of the exposed sediments. Only a small proportion of these fossils are hominins.

Australopithecus anamensis

The earliest *Australopithecus* fossils currently known date from between 3.8 and 4.2 million years ago. These belong to a species called *Australopithecus anamensis*, which was established by palaeoanthropologist Meave Leakey and colleagues in 1995 when they discovered a series of fossils from Kanapoi and Allia Bay near Lake Turkana in Kenya. The name is derived from anam meaning 'lake' in the Turkana language. Remarkably the first hominin fossil from Kanapoi had been

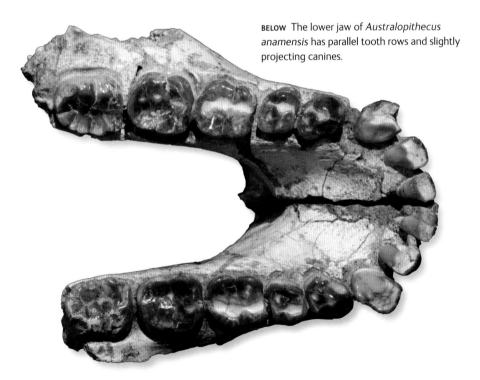

BELOW The lower jaw of *Australopithecus anamensis* has parallel tooth rows and slightly projecting canines.

found almost 30 years earlier, in 1965, but never ascribed to a species. This was the lower part of an arm bone, which intrigued scientists for many years and is now included within *Au. anamensis*.

The *Au. anamensis* dentition exhibits a combination of primitive and derived features. The canines project only slightly beyond the neighbouring teeth but they are broader at the base than those of more recent *Australopithecus* species such as *Au. afarensis*. The lower front premolar is more primitive with a more elongated shape than later hominins and it has a single central cusp. The front teeth are fairly large compared to more recent hominins and the dental arcade is U-shaped with the cheek teeth forming parallel rows.

In 2019 scientists announced the discovery of a remarkably complete 3.8-million-year-old *Au. anamensis* cranium from the Woranso-Mille area of Ethiopia. The cranium was found by a shepherd buried in sand that had been deposited on the shores of an ancient lake. The new fossil revealed features of the cranial vault and face of *Au. anamensis* for the first time. The braincase was long and narrow and less constricted behind the eye orbits than in some more recent australopiths. The facial skeleton revealed forward-projecting cheekbones, a deep palate and a projecting face. The new find demonstrated that *Au. anamensis* and the more recent *Au. afarensis* (see pp. 52–61) could not be part of a single evolving lineage as previously thought, because *Au. anamensis* exhibited some derived anatomical features that are absent in *Au. afarensis*. New dating evidence demonstrated that the two species coexisted in eastern Africa for around 100,000 years between about 3.9 and 3.8 million ago.

RIGHT The upper part of the tibia (shin bone) of *Au. anamensis* reveals that this species walked on two legs. The tibial plateau, where the tibia and femur (thigh bone) connect, is relatively large, which suggests that more weight was transferred across the knee joint, and is consistent with upright walking.

ABOVE The nearly complete cranium of *Au. anamenis* from Woranso-Mille revealed the face of this species for the first time.

Several postcranial elements are represented in the *Au. anamensis* fossil assemblage. Among these are both ends, but not the middle, of a tibia (shin bone) from Kanapoi. This bone provides vital clues about body size and the way in which this species moved around. The joint surface of the lower part of the tibia is perpendicular to the shaft of the tibia. This implies that the lower half of the leg was angled directly downwards from knee to ankle, as in modern humans. In chimpanzees the tibia is angled outwards from the ankle because the knees are positioned further apart than in humans, giving a bow-legged appearance. The reorientation of the long axis of the tibia is a good indication that *Au. anamensis* walked on two legs. Scientists estimate that *Au. anamnesis* weighed about 46 kg (101 lb), based on the size of the upper part of the tibia.

The shape of the capitate (one of the wrist bones) suggests that *Au. anamensis* may not have been able to produce a precise thumb to fingertip grip. This is similar to living apes and differs from later hominins such as *Au. afarensis*, as well as some fossil apes.

The carbon isotope values in teeth from *Au. anamensis* are similar to those of forest-dwelling chimpanzees and gorillas, suggesting that they ate fruits and other edible parts of trees, bushes and shrubs, and very few savannah foods such as grasses or sedges. Many of the hominin fossils from Kanapoi show damage caused by carnivores, suggesting that the bones were gnawed by predators or scavengers.

Australopithecus afarensis

In 1974, scientists recovered a skeleton of a small female hominin in the Afar region of Ethiopia that was more complete than any other early hominin known at the time. She was nicknamed Lucy in honour of the Beatles' hit *Lucy in the Sky with Diamonds*, and was one of the first hominin fossils to become a household name. Lucy was part of a spectacular assemblage of hominin fossils found at Hadar in Ethiopia between 1973 and 1976. Don Johanson, Tim White and Yves Coppens established a new species, *Australopithecus afarensis*, in 1978 to accommodate the Ethiopian fossils and similar fossils from slightly older deposits at Laetoli in Tanzania.

Australopithecus afarensis is now one of the best represented species in the hominin fossil record, so its chronological and geographical distributions, as well as its morphological variation, are well documented. Fossils from *Au. afarensis* have been found at sites in Ethiopia, Kenya and Tanzania and are dated between 3 and 3.9 million years old.

Australopithecus afarensis had a prognathic (projecting) face and a slightly domed cranial vault with a small braincase. Its cranial capacity ranged from about 385 cm^3 to 550 cm^3. Some specimens of *Au. afarensis* have a small sagittal crest towards the back of the skull and a temporonuchal crest. The sagittal crest is a bony ridge that runs along the midline of the top of the skull that increases the area available for attachment of one of the main chewing muscles. The temporonuchal crest forms at the back of the skull where the attachments of chewing muscles are close to those of the neck muscles. Both of these crests indicate powerful muscle attachments.

RIGHT Lucy was a competent biped but her shoulders and arms were well adapted for moving around in the trees.

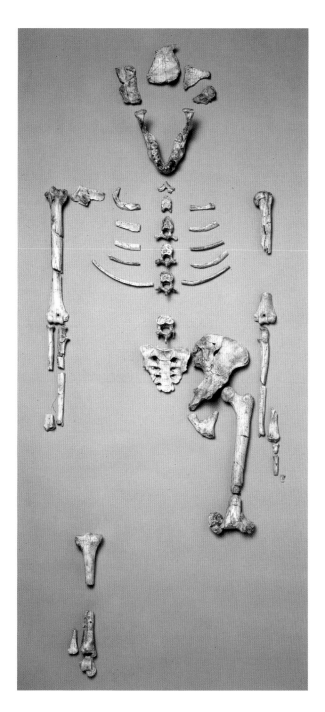

LEFT The partial skeleton of Lucy was the most complete early hominin known to scientists when she was discovered in 1974. Parts of the skull and pelvis are present, together with bones from both arms and legs and some of the vertebrae and ribs.

Some *Au. afarensis* have tooth rows that diverge slightly towards the back, forming a dental arcade that is neither parallel-sided as in modern apes nor parabolic as in humans. The canine teeth of *Au. afarensis* are much smaller than those of a chimpanzee, and they are less broad and differently shaped to those of the earlier *Au. anamensis*. The canine premolar honing complex has been completely lost.

The postcranial skeleton reveals an accomplished biped although *Au. afarensis* may not have walked in exactly the same way as modern humans today or been capable of walking long distances efficiently. Anatomical features associated with upright walking are present in the vertebral column, pelvis, legs and feet. For example, the pelvis of *Au. afarensis* is very broad and has short and flaring iliac blades (see p.29). The femur (thigh bone) is angled inwards towards the knee, so that the centre of gravity falls directly above the foot. As with *Au. anamensis*, the long axis of the tibia (shinbone) is perpendicular to the joint surface between the lower part of the tibia and the ankle. The big toe is in line with the other toes and the toe bones are relatively short like those of humans, but curved like those of an ape.

BELOW *Australopithecus afarensis* and other early hominins may have built sleeping platforms in the trees, like these orang-utans.

Lucy and her species also retained some adaptations for arboreal climbing and suspension. These features are seen in the shoulders, arms, wrists and hands. *Au. afarensis* may have been foraging in the tree canopy as well as on the ground, and probably retreated to the trees at night to avoid predators and for a good night's sleep. Chimpanzees and other apes are known to build nesting platforms in tree canopies. Studies have shown that apes sleep more soundly than monkeys, who attempt to sleep while balanced in an upright position on tree branches or rocky ledges.

Various lines of evidence suggest that *Au. afarensis* occupied a slightly different dietary niche to that of earlier hominins. Carbon isotope values in tooth enamel reveal that *Au. afarensis* is the earliest hominin species showing evidence for a more diverse diet that included savannah-based foods such as sedges or grasses as well as a more traditional diet based on fruits and leaves from trees and shrubs. Some of the anatomical differences between *Au. anamensis* and *Au. afarensis* are adaptations for heavy chewing suggesting that there was a change in diet towards foods that were harder or tougher over time. The front teeth of *Au. afarensis* are less worn down than those of *Au. anamensis* indicating that they had a less important role in food acquisition and processing.

Another small child

Eighty years after the discovery of the Taung child from South Africa, another *Australopithecus* child took centre stage. In 2006, palaeoanthropologist Zeresenay Alemseged and his team found an almost complete skeleton of a tiny *Au. afarensis* child eroding out of a hillside at Dikika in Ethiopia. It was just a few kilometres away from the site in Hadar where Lucy had been unearthed.

Initially only the face of the child was exposed on a dusty slope, but much of the skeleton was preserved below the surface. The process of carefully extracting the tiny bones from the sandstone matrix to reveal details of its preservation and anatomy took more than five years. As the bones lacked any marks caused by predators or scavengers, the child must have died naturally or in an accident and been quickly buried. One possibility is that she was swept away and drowned in a flash flood, and that her body sank into sediments transported by the river.

The skeleton was so complete that it preserved elements that were previously unknown in *Au. afarensis*. The Dikika child had a full set of milk teeth and scientists made CT-scans of the skull allowing them to view the permanent teeth

ABOVE The skeleton of a tiny *Australopithecus afarensis* child, nicknamed Selam, was discovered eroding out of 3.3 million year old deposits at Dikika in Ethiopia.

developing in the jaws. This revealed that a state of dental development that is similar to that of a three-year-old chimpanzee. The permanent canines were small compared to other *Au. afarensis* specimens, suggesting that the child was female. A more recent study of the Dikika child using synchrotron CT scans revealed further insights into her growth and development. Scientists were able to determine a more precise age at death of only 2.4 years by counting microscopic growth lines inside her teeth. The child had a brain size of only 275 ml, pointing towards an extended period of brain growth compared to chimpanzees, and a longer period of reliance on caregivers.

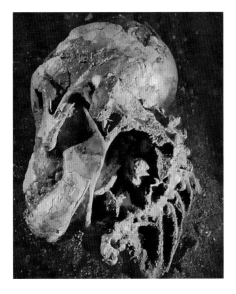

ABOVE Reconstruction of how the *Australopithecus afarensis* child known as Selam may have looked shortly before her death.

LEFT Described as 'a precious little bundle', Selam may have been swept away by a flash flood and rapidly buried by sediment, protecting her body from scavengers.

With shoulder blades most similar to those of a gorilla and long curved finger bones, the little girl would have been an adept climber, but as with adults of her species, her lower body was well adapted for upright walking. She had big toes that were able to grip. This would have helped her to hold onto her mother when carried and to move more easily in the trees to avoid predators or to find food or a safe place to sleep.

Laetoli footprints

The site of Laetoli in Tanzania preserves the earliest known hominin footprints. Nearly 3.7 million years ago, a volcanic eruption covered the surrounding landscape with a layer of fine volcanic ash. A light sprinkling of rain fell on the ash shortly after it had settled, creating a surface like wet cement. The animals that chanced to walk across this landscape before the ashy surface had time to harden left behind their footprints. There are more than 20 species represented, including rhinoceros, giraffe, hyena, baboon, antelope and various birds. Among the trails of prints there are some that were made by hominins. Further volcanic eruptions covered the footprints, preserving them like a snapshot in time.

The first animal prints were spotted at the site in 1976, and two years later palaeoanthropologist Mary Leakey excavated the trail of hominin footprints. The trail extends for nearly 27 m (88½ ft) and is made up of about 70 footprints. This remarkable find represents a rare fossilized record of hominin behaviour and provides clues about the anatomy of the species that made the prints. Scientists attributed the hominin footprints to *Au. afarensis* as fossils from this species were found in sediments from the same date and location and, at the time of discovery, no other hominin was known to have been living in the area during that period. Although other bipedal hominins are now known to have existed alongside *Au. afarensis* in eastern Africa, *Au. afarensis* still seems the most likely candidate for the footprints at Laetoli since no fossils from other hominin species have been recovered at the site.

The footprints were produced by a group of individuals, with at least one walking behind and stepping into the footprints already made by a bigger individual. The close spacing of the hominin footprints implies that the hominins who made them had short legs. The shape of the prints indicates a modern-looking foot with an arch and with the big toe aligned with the other toes. The foot strike resembles that of modern humans, with the heel touching the ground first and weight transferring to the ball of the foot before the toes push the foot off the ground at the end of each stride. Some scientists interpret this as evidence of a fully modern form of bipedal walking, but others have suggested that the leg may have been slightly more bent at the knee than that of modern humans at the point when the heel strikes the ground.

Forty years after the original discovery, Tanzanian scientists uncovered another set of footprints just 150 m (492 ft) south of the original trail. The new tracks were made by two hominins walking in the same direction as the original group. One of these hominins produced a much larger footprint than any of those previously recorded and must have been made by a taller and heavier individual. The two trails may have been a single social group, either an adult walking with a group of children or a large-bodied male walking with several smaller-bodied females and children.

OPPOSITE The footprints preserved at Laetoli in Tanzania include those of a small group of hominins who walked across ash from a volcanic eruption nearly 3.7 million years ago. The tracks were reburied after they were recorded to help preserve them.

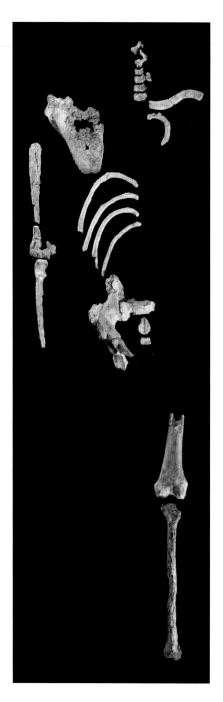

LEFT Nearly half of the skeleton known as Kadanuumuu, meaning Big Man, was recovered from the Woranso-Mille study area of Ethiopia but the feet, hands and skull were missing.

Kadanuumuu

More than 30 years after the partial skeleton of Lucy was found, scientists discovered another similarly complete skeleton of *Au. afarensis* at a site in the Woranso-Mille study area, about 50 km (31 miles) north of Hadar. At 3.58 million years old, the new skeleton is nearly 400,000 years earlier than Lucy and dates from a similar period to the footprints at Laetoli. The new skeleton is also considerably taller than Lucy, with an estimated height of 1.5–1.7 m (5–5½ ft) compared to Lucy's 1.05 m (3½ ft) stature. The fossil skeleton was nicknamed Kadanuumuu, meaning 'Big Man' in the Afar language.

Nearly half of Kadanuumuu's skeleton was recovered and some bones such as the clavicle and scapula are exceptionally well preserved. The difference in body size between Lucy and Kadanuumuu is consistent with other evidence for sexual dimorphism in this species. Sexual dimorphism in living primates is often expressed as a ratio of average male body-weight to average female body-weight. Humans are not very dimorphic and neither are chimpanzees. Male chimpanzees and bonobos are about 30% bigger than females and in humans the difference between males and females is even lower. Gorillas exhibit high sexual dimorphism in body mass, with males weighing about twice as much as females on average. High sexual dimorphism is linked to social systems in which males are competitive towards other males and may have multiple mates. Since we cannot weigh fossil hominins, scientists use measurements of the skeleton to estimate body mass. The smallest adults from *Au. afarensis* probably weighed about 25 kg (55 lb) and the largest about 64 kg (141 lb), and this broad range of body sizes within a single species points to high sexual dimorphism.

Complete and partial skeletons are valuable because they are conclusive evidence that a particular set of bones and teeth belong together, representing an example of a single species at a single moment in time. Fragmentary fossils can provide a wealth of anatomical information, for example part of a leg bone can tell us whether a species was bipedal and a handful of bones can reveal tool making capabilities; but if more than one hominin species occurs at a site or at nearby sites from the same period it is not always obvious which species they belong to. Specimens that are as complete as those of Lucy, Kadanuumuu and the Dikika child can be used to build up a comparative skeletal atlas for the species they represent and help the identification of fossil finds in the future. Nevertheless, some scientists have questioned whether Kadanuumuu can reliably be assigned to *Au. afarensis* since there are no skull or dental parts associated with the skeleton so it is not possible to make a direct comparison with the fossil jaw that formed part of original description of *Au. afarensis*.

Australopithecus bahrelghazali

Hominin 1 from site 12 at Koro Toro in Chad is known to scientists as KT-12/H1. It consists of nothing more than the front part of a lower jaw with one incisor, both canines and four premolars. Though fragmentary, it was a remarkable discovery because it represents the first evidence for *Australopithecus* in central Africa – all other specimens had been found in sites close to the Rift Valley or from South Africa. In 1996, palaeontologist Michel Brunet and colleagues assigned KT-12/H1 to a new species, *Australopithecus bahrelghazali*, meaning 'river of the gazelles' in Arabic and named after the place of discovery.

Australopithecus bahrelghazali lived on the ancient shoreline of Lake Chad and dates from about 3.6 million years ago. Other animals found at the same site, such as turtles, elephants, antelopes and rhinoceroses, indicate a lakeside environment with a mosaic of open grassland, wooded savannah and gallery forest. Carbon isotope values in the teeth reveal that *Au. bahrelghazali* had a different diet to other early hominins that was made up almost entirely of C4 foods such as tropical grasses and sedges (see p.16). One possibility is that this species specialized in eating sedges growing on the lakeside.

Not much is known about the anatomy of *Au. bahrelghazali* because so far only a part of a lower jaw and one tooth from an upper jaw have been discovered. Some scientists have suggested that the fossils assigned to *Au. bahrelghazali* could belong to *Au. afarensis* or to one of the other hominins that lived about 3.5 million years ago. The fossils are important because they demonstrate that hominins were in central Africa during this period. Their unusual diet also provides tantalizing evidence of a different way of life.

Kenyanthropus platyops

Kenyanthropus platyops means 'flat faced man from Kenya' and is the name given by Meave Leakey and colleagues to a new hominin genus and species described in 2001. The fossils were found at the site of Lomekwi, west of Lake Turkana in Kenya, and date from between 3.3 and 3.5 million years ago. The most complete *Kenyanthropus* fossil is a nearly complete cranium that displays a unique combination of anatomical features. It has tall and forward-placed cheek bones and the part of the face below the nose does not project forward, giving the appearance of a flat face. These features distinguish *K. platyops* from *Au. afarensis*, the best known hominin from eastern Africa in the same time period.

The facial architecture of *K. platyops* exhibits striking similarities with a more recent cranium from Koobi Fora in Kenya belonging to *Homo rudolfensis* (see p. 85). These features may have arisen independently or may imply a close evolutionary relationship. The cranial capacity of *K. platyops* could not be measured precisely due to the deformation of the cranial bones but its brain size appears to be similar to that of *Australopithecus* and was certainly much smaller than *H. rudolfensis*.

Carbon isotope analysis of fossil hominin teeth has revealed that *K. platyops* was one of the earliest hominin species to expand its diet to include foods from open savannah environments alongside those from forest and woodland. These newly exploited foods might have included leaves, seeds, roots and tubers from grasses or sedges or potentially other animals or insects that ate those foods. The molars, which are used for crushing and grinding food have thick enamel, but are small compared to most *Australopithecus* species.

ABOVE *Kenyanthropus platyops* has a unique combination of anatomical features and its relationship to other hominins is not yet known.

Put your best foot forward

The recent discovery of an intriguing set of foot bones from the Burtele locality at Woranso-Mille in Ethiopia is further evidence for the presence of more than one species of hominin in eastern Africa 3.4 million years ago. The anatomical features of the Burtele foot are different from *Au. afarensis* foot bones, demonstrating that there were at least two different ways of walking upright among hominins living between 3 and 4 million years ago. The foot bones from Burtele share some features with those of *Ar. ramidus*, known to have been living in the Afar region almost exactly a million years earlier (see p. 27). Both sets of foot bones reveal a divergent big toe capable of grasping branches. The Burtele foot bones were found at the same locality as partial jaw bones assigned to *Au. deyiremeda*, leading to speculation that they might have belonged to the same hominin species.

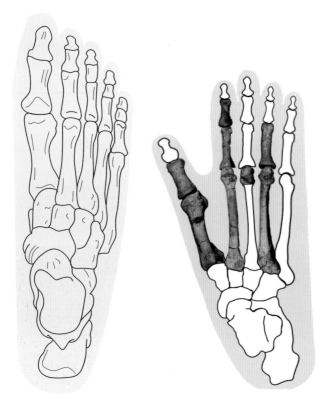

LEFT The bones of the big toe are aligned with the other bones in the modern human foot (far left) but diverge away from the other toes in the fossil foot from Burtele (left). The Burtele foot reveals that a hominin capable of walking on two legs but still well adapted to tree-climbing was living in Ethiopia about 3.4 million years ago.

Join the crowd

The most recent addition to the increasingly crowded group of species living at about 3.5 million years ago is *Australopithecus deyiremeda*. The new species was established by palaeoanthropologist Yohannes Haile-Selassie and colleagues in 2015 from fossils collected in the Woranso-Mille area in the Afar region of Ethiopia. The species name means 'close relative' in the language of the Afar people. The new species is represented by a partial upper jaw and parts of two lower jawbones that date from between 3.3 and 3.5 million years ago. Teeth belonging to *Au. deyiremeda* are smaller than those of most *Au. afarensis* specimens and have differently shaped roots, and its canines are particularly small.

Australopithecus deyiremeda overlaps in time with *Au. afarensis*, *Au. bahrelghazali* and *K. platyops* as well as with hominin species found in South Africa. Fossils assigned to *Au. bahrelghazali*, *K. platyops* and *Au. deyiremeda* can be directly compared to *Au. afarensis* because this species is well represented in the fossil record. In each case the scientists who named the new species consider them to

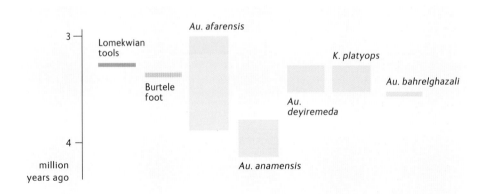

ABOVE Recent fossil discoveries have revealed that a diverse array of hominin species inhabited eastern Africa between 4 and 3 million years ago, including several species of *Australopithecus*, *Kenyanthropus platyops*, and an unassigned foot fossil from Burtele, Ethiopia. At least one of these species was making and using stone tools such as those produced at Lomekwi.

RIGHT A partial lower jaw bone from *Au. deyiremeda*.

fall outside the known range of variation of *Au. afarensis*. Comparisons between the newer species themselves are constrained by the small numbers of fossils collected. The species identity of isolated or less complete fossils previously assigned to *Au. afarensis* may need to be reconsidered now that other hominin species are known to have co-existed.

The presence of multiple hominin species in eastern Africa between 3.3 and 3.9 million years ago raises questions about how they divided up the landscape and available resources. They almost certainly had different food preferences or relied on different fall-back foods in times of scarcity. Carbon isotope values reveal a marked shift in hominin diet between about 3.5 and 4 million years ago. At that time most species expanded their diet to include C4 foods such as grasses and sedges. This expansion and diversification of diet would have enabled species such as *Au. afarensis* and *K. platyops* to live in a broader range of habitats. This shift coincides with other evidence that hominins were occupying more open wooded or savannah environments. *Au. bahrelghazali* from Chad appears to have had a different diet to the other species, relying almost exclusively on C4 foods. This may reflect opportunist exploitation of a particularly rich lakeside environment rather than a specialized dietary adaptation.

Australopithecus garhi

In 1999, palaeontologist Berhane Asfaw and colleagues announced the discovery of a previously unknown hominin from Bouri in the Middle Awash area of Ethiopia. They named their new species *Australopithecus garhi*. The word garhi means 'surprise' in the Afar language. The fossils date from about 2.5 million years ago. *Au. garhi* lived half a million years after the last known *Au. afarensis* and is the most recent evidence of *Australopithecus* in East Africa. It overlaps in time with *Paranthropus* (see p. 71) and early *Homo* (see p. 83) in eastern Africa and South Africa and with *Au. africanus* in South Africa.

The most complete fossil is a partial cranium, which reveals that *Au. garhi* had a projecting face and a small sagittal crest running along the midline of the skull. The upper jaw retains an almost complete set of teeth, revealing large chewing teeth and broad canines. Its cranial capacity has been estimated at 450 cm^3, which is comparable to *Au. afarensis* and *Au. africanus*. A partial skeleton found about 300 m (984 ft) from the fossil skull and other fossils found in the same locality may be associated with *Au. garhi*. The limb proportions are

ABOVE *Australopithecus garhi* from Ethiopia demonstrates that australopiths were still present in eastern Africa about 2.5 million years ago. The reconstruction is speculative as the bones that connect the upper jaw to the cranial vault are not preserved in this fossil.

unexpected for *Australopithecus*. The skeleton exhibits a relatively elongated femur suggesting habitual walking, but the forearm remains relatively long. The size variation in the upper arm bones may imply sexual dimorphism, with males larger than females.

Cut marks and other traces on the fossils of antelopes and other large mammals at Bouri are currently the earliest widely accepted direct evidence for butchery. The modified bones reveal that the animals were disarticulated and defleshed using stone tools and that long bones were broken open to gain access to marrow. These modifications are evidence that at least one species of hominin was hunting large animals or scavenging from the carcasses of animals killed by other predators, such as lions or leopards.

THE EARLIEST TOOL-MAKERS

Modern humans have an unrivalled capacity to manufacture tools and use them to transform the environment in which they live, and to produce other items for practical and symbolic purposes. Although the earliest known evidence for tool-making is less than 3.5 million years old, it is likely that all hominins made use of tools to some extent. Our closest relatives, the chimpanzees, have been observed making tools from grass stems for 'termite fishing' and using rocks as tools to process food, further closing the behavioural gap between them and us. Other apes and monkeys use tools made from plants and a few use crude stone-based technologies, for example to crack open nuts.

Direct evidence for the use and manufacture of stone tools comes from the presence of the tools at a fossil site or from evidence of 'cut marks' on fossil animal bones. The first known stone tools, dating from about 3.3 million years ago and made from large pieces of volcanic rock, come from the Kenyan site of Lomekwi. The Lomekwian stone tool industry is technologically diverse and includes cores, flakes and potential anvils. From about 2.6 million years ago, there are smaller-sized pebble tool industries, known as 'Oldowan', after the site of Oldupai Gorge in Tanzania, and the earliest known Oldowan stone tools are from Gona in the Afar region of Ethiopia. These early tools are simple, with limited working and an unstandardized shape. Cut marks are the characteristic trace marks left by stone tools when they are used to butcher the carcasses of hunted, or more likely in the earliest stages, scavenged, animals. The earliest undisputed evidence for butchery dates from 2.5 million years ago and was found at Bouri, not far from Gona, where cut marks and other damage to the bones of large mammals reveal that the carcasses were butchered to extract meat and marrow.

For a long time it was assumed that only members of the genus *Homo* were capable of manufacturing artefacts, but recent discoveries have revealed that the earliest stone tools pre-date the earliest evidence of the genus *Homo* by half a million years, implying that earlier hominins were also able to make stone tools. Scientists have become more cautious about assuming an association between a particular species and tools or cut-marked bones found at the same site, as there is increasing evidence for species diversity at every stage of hominin evolution. Some sites contain fossils from two or more hominin species, and at other sites there may have been more than one species of hominin present, even if their fossils have

LEFT The earliest known stone tools belong to the technologically diverse Lomekwian industry from Kenya.

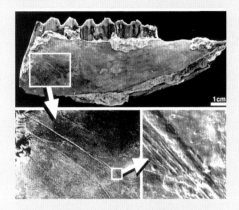

ABOVE Animal bones from Bouri in Ethiopia reveal that hominins were already butchering animal carcasses by 2.5 million years ago. The images illustrate cut marks on a lower jaw made by a stone tool.

not been found. The species responsible for making the stone tools at Lomekwi has not yet been conclusively identified, but *Kenyanthropus platyops*, which is the only hominin species known from this site around 3.3 million years ago, is the most likely candidate. Similarly, *Australopithecus garhi* is currently the species most likely to have made the stone tools at Gona as it is is known from contemporary deposits at the nearby site of Bouri.

Tools made from stone are most likely to survive in the archaeological record but tools can be produced from a variety of materials. Wood, bamboo and animal hides are not as durable as stone, and are only preserved under special circumstances so much evidence of past human creativity in materials other than stone has been lost. It is quite possible that even before stone tools were first made, there was a phase of tool-making and tool-using we cannot recognize that involved the use of leaves, wood or bones.

Occasionally tools manufactured from other material do survive in the archaeological record particularly in more recent times. Several hominin sites in southern Africa, including Sterkfontein, Swartkrans and Drimolen, have yielded bone tools dated to between 1 and 2 million years ago. Analysis of the wear patterns on these tools reveals that they might have been used for 'termite fishing', or to dig up underground tubers or process fruit. Fossils of *Paranthropus* and early *Homo* occur at most of the sites where bone tools have been found so for now, attempts to identify the species that produced and used the tools are often inconclusive.

Scientists have also looked at the anatomy of the hand to infer tool-making abilities. Adaptations in the hand and wrist bones can reveal whether hominins were capable of manipulating objects with the force and precision required for making and using stone tools effectively. Scientists have identified three key adaptions of the human hand and a suite of anatomical features that can be used to detect these capabilities. The first, precision handling, is the ability to rotate and manipulate objects using the thumb and fingertips of one hand; humans do this when we pick up and twiddle a pen. The second, a precision grip, describes the ability to press the thumb and fingertips together forcefully; and the third, a power squeeze grip is the ability to grip an object diagonally across the palm of the hand with the fingers and stabilize it with the thumb. Reasonably complete hands are rare in the hominin fossil record, but those that have been found reveal that that some species of *Australopithecus* and *Paranthropus*, as well as *Homo naledi* (see p. 101), had hands that were well adapted for both controlled manipulation of objects and climbing.

CHAPTER 4

Paranthropus

*P*ARANTHROPUS IS THE NAME used for a group of species from eastern Africa and South Africa that share a distinctive suite of dental and cranial adaptations. They are characterized by massive cheek teeth and skulls that were capable of generating and withstanding powerful chewing forces. Their flared cheek bones are one of the most striking features. These extend far forward helping to create the impression of a dish-shaped face. *Paranthropus* species have enlarged molars with very thick enamel and the premolars are larger and shaped more like molars (molarized) to increase their chewing capabilities. In contrast, their incisors and canines are small and probably did not serve an important function during food acquisition and processing.

The *Paranthropus* group originated at least 2.6 million years ago and is thought to have gone extinct sometime after 1.3 million years ago. There are various theories as to why they became extinct. They may have been so specialized in their diet that they were not able to adapt to changing environments. Alternatively, they may have been unable to compete with other species such as baboons or other hominins that exploited similar resources more efficiently.

OPPOSITE Drimolen is an archaeological site in the Cradle of Humankind. The Main Quarry has been excavated since the 1990s and has yielded fossils of *Paranthropus robustus* and early *Homo*, as well as bone tools made by one of these hominins. The fossil assemblage is dated between approximately 2 and 1.5 million years ago.

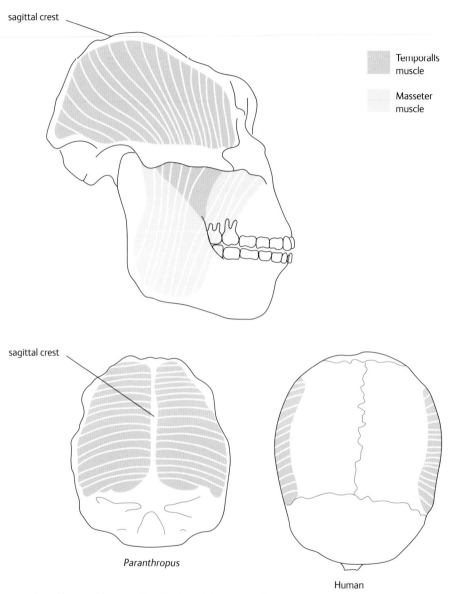

ABOVE *Paranthropus* (above and top) had much larger areas for attachment of the chewing muscles than modern humans (right). The masseter muscle attaches the lower jaw to the cheek bone, and is used to open and close the mouth. The temporalis muscle attaches the lower jaw to the sagittal crest in *Paranthropus* and to the sides of the braincase in modern humans, and is used for crushing and grinding foods.

The use of the genus name *Paranthropus* implies that all the species in this group shared a unique common ancestor and are more closely related to each other than to species in another genus such as *Australopithecus*. However, some scientists think that shared features of the *Paranthropus* group could have evolved independently in eastern Africa and South Africa. This is because the features relate to a single functional complex (in this case powerful chewing), which might have evolved more than once under similar selective pressures. Scientists who take this view refer to these hominins using the informal group name 'robust australopiths', which does not involve any assumptions about their evolutionary relationships.

Paranthropus aethiopicus

In 1967, scientists working at Omo in Ethiopia recovered a toothless lower jaw from sediments dating from about 2.6 million years ago. The size of the tooth sockets and surviving tooth roots suggested that the large molars at the back of the jaw were accompanied by relatively smaller incisors and canines at the front of the jaw. Since its early date and anatomy precluded it from belonging to any other hominin species known at the time, palaeontologists Camille Arambourg and Yves Coppens described the specimen as a new species, *Paraustralopithecus aethiopicus*.

Almost two decades later, an almost complete and equally toothless cranium was found at West Turkana in Kenya. The fossil became known as the Black Skull because of its distinctive blue-black appearance (due to the uptake of manganese during fossilization). The skull had a small brain case, with an estimated volume of 410 cm^3. The combination of a small braincase and massive chewing muscles has resulted in the formation of an impressive set of cranial crests to enlarge the areas available for muscle attachment. The sagittal crest extends all the way from the top of the skull to the nuchal crest, a bony ridge at the back of the skull that provides an additional attachment area for the neck muscles.

Most scientists now include the Black Skull and Omo mandible, together with a small number of other fossils from the same period, as *Paranthropus aethiopicus*. So far no fossils from the postcranial skeleton have been assigned to this species. *Paranthropus aethiopicus* is the earliest species in the *Paranthropus* group, dating from 2.3 to 2.6 million years ago, and it lacks some of the shared anatomical features of the more recent *Paranthropus* species. Compared to the later species, it has smaller cheek teeth and larger incisors, and the lower part of the face projects further forward.

ABOVE The evocatively named Black Skull from West Turkana in Kenya is the most complete fossil belonging to *Paranthropus aethiopicus*. The striking colour is caused by the uptake of manganese from the surrounding soil during fossilization.

Paranthropus robustus

The first fossils of *Paranthropus robustus* were discovered in 1938 by a schoolboy, Gert Terblanche, but their significance was recognized by Robert Broom, who occasionally purchased hominin fossils from Mr Barlow, the quarry manager at Sterkfontein. On one occasion he was offered a small piece of upper jaw that looked different from hominins known at the time. Broom was intrigued and he noticed that two teeth had been freshly broken off the jaw. On asking Mr Barlow if he had the teeth, the quarry manager revealed that he had in fact bought the fossil from a local schoolboy, Gert Terblanche. Broom found Terblanche at his school whereupon the boy pulled a handful of teeth from his pocket. He later took Broom to an eroded cave deposit at Kromdraai, a few kilometres from Sterkfontein, where he had found the fossils and hidden the lower jaw. Over the next few days, Broom recovered more pieces from the cranium. He recognized that the skull was more heavily built ('robust') than any other hominin known at the time. He announced the new skull in 1938 and assigned it to a new genus and species, *Paranthropus robustus*.

Paranthropus robustus is the only hyper-robust early hominin from South Africa and it is now known from six collapsed cave sites, all within a few kilometres of Kromdraai. *P. robustus* lived between about 1.4 and 2.2 million years ago. The species is characterized by massive chewing teeth with thick enamel, molarized premolars and diminutive incisors and canines. The cheek bones are large and shifted forwards so they project beyond the nose and upper face, producing a dish-shaped face. The flaring cheek bones also provide space for an enlarged temporalis muscle, which is one of the main muscles used during chewing. Some of the skulls have a sagittal crest located at the top of the skull to increase the area of attachment for the temporalis muscles.

The distinctive morphology of the dentition and skull are adaptions for powerful chewing, suggesting ability to crack open hard objects or to grind tough and fibrous plant foods. The carbon isotope values in tooth enamel from *P. robustus* indicate a diet that included forest and savannah-based foods. More than 100 bone tools have been identified at sites including Sterkfontein, Swartkrans Cave, Drimolen and Cooper's D. Fossils belonging to both *P. robustus* and early *Homo* have been recorded at most of these sites, so it is unclear which species used the tools, but so far Cooper's D only has only yielded fossils of *P. robustus* suggesting that this species had the cognitive and manipulated abilities required to make and use tools. Experimental studies and use

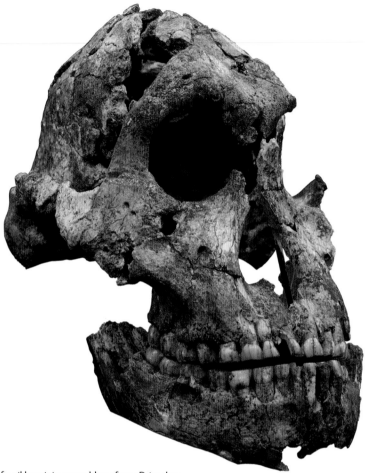

ABOVE The fossil hominin assemblage from Drimolen includes an almost complete adult skull of *Paranthropus robustus*. The skull is thought to be female because it lacks the sagittal crest which is found in adult males of this species.

wear indicate that the bone tools were used to extract food resources, such as digging for tubers or termite harvesting.

Several parts of the postcranial skeleton of *P. robustus* are known. *Paranthropus robustus* is thought to have weighed between about 24 kg (53 lb) and 43 kg (95 lb) when fully grown, meaning that the largest adults were almost twice as heavy as the smallest. There is also variation in the size and shape of the skulls, and only the larger skulls display a pronounced sagittal crest. This variation in size and shape is consistent with sexual dimorphism, and suggests a social system where males were competing with each other to take control of a group of females. The cranial evidence reveals that *P. robustus* has a cranial capacity of about 530 cm^3, implying that this species had a slightly larger brain than most living apes and *Australopithecus*.

Tell-tale marks on the bones of *P. robustus* have revealed how some of the fossil assemblages accumulated. At Swartkrans, another collapsed cave system not far from Kromdraai, there are depressions on the back of the skull of a young *P. robustus* that exactly match the size and spacing of punctures made by the lower canines of a leopard. Leopards regularly stash their prey in trees to keep it out of reach of scavengers, so it is thought that a leopard killed the young hominin and dragged the corpse to a branch overhanging the site where the fossil was found. Puncture marks and damage on other bones at Swartkrans reveal that several other predators and scavengers contributed to the formation of the fossil assemblage, including lions, spotted hyenas and large dogs.

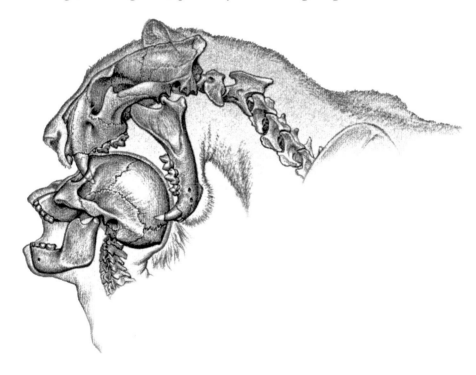

ABOVE Indentations on the back of a skull of a young *Paranthropus robustus* found at Swartkrans suggest it was seized by a leopard.

Paranthropus boisei

The famous palaeoanthropologist Louis Leakey first started to collect stone tools when he was a boy living in Kabete, a village near Nairobi in Kenya. Less than a day after his arrival at Oldupai Gorge in Tanzania in 1931 he picked up a stone tool and won a ten pound wager with another member of his expedition who was convinced that there were no stone tools at this site. Over the next three decades many thousands of stone tools and numerous animal bones were found at more than 30 sites at Oldupai Gorge, but fossils of the hominins that made those tools remained elusive. This changed in 1955, and within a few excavation seasons fossils from not just one but three clearly distinguishable hominin species were unearthed at Oldupai. The first major discovery was made on 17 July 1959 by Louis Leakey's wife, Mary Leakey, who spotted a

BELOW Oldupai Gorge in Tanzania has yielded fossils of *Paranthropus boisei* and *Homo habilis*, together with thousands of stone tools and animal bones with cut marks.

piece of skull and brushed back the sediment to reveal two hominin teeth. The massive teeth showed a clear resemblance to *Paranthropus robustus* and the cranium had a small brain case and a well-developed sagittal crest. This opened a conundrum for Louis Leakey, who believed that only a larger-brained direct human ancestor could have made the stone tools at Oldupai. Louis Leakey described the skull as a new genus and species, *Zinjanthropus boisei*, and initially accepted albeit reluctantly that this small-brained hominin must have been an accomplished tool-maker. Subsequent discoveries of fossils belonging to the genus *Homo* at Oldupai reopened that debate and many scientists now believe that the earliest tools currently known at Oldupai Gorge were made by *Homo habilis* (see p. 85). The formal number for the first Oldupai cranium is OH5, but it has been known by a series of nicknames including Dear Boy, Nutcracker Man and Zinj, and is now placed in the genus *Paranthropus*.

Paranthropus boisei is now known from East African sites ranging from southern Ethiopia to Malawi, dated between 1.3 and 2.3 million years ago, and is the most specialized species belonging to *Paranthropus*. It has tiny peg-like front teeth, cheek teeth with extremely thick enamel and chewing surfaces that are up to four times as large as those of modern humans. Despite this formidable chewing apparatus, the teeth wore quickly so the softer dentine that underlies the hard enamel was exposed at an early age.

The highly specialized teeth and cranial architecture of *P. boisei* led scientists to assume that its diet included hard and tough objects that required considerable mechanical force to break down. Surprisingly, analysis of microwear on *P. boisei* teeth did not reveal any evidence that they had been consuming hard or tough foods. Stable isotope analysis has revealed that the diet of *P. boisei* was dominated by savannah-based C4 foods such as grasses or sedges. *P. boisei* was uniquely specialized in this way, relying more heavily than any other hominin on these types of food. Both sedges and grasses are low-quality foods with a low protein and high fibre content. This suggests that the specialized cranial and dental morphology of *P. boisei* may have been an adaptation for processing large quantities of low quality foods rather than foods that were mechanically challenging. The enormous flat-wearing cheek teeth would have been needed to grind up poor-quality foods.

Until recently, very little was known about the anatomy of *P. boisei* from the neck down. Isolated post cranial bones cannot reliably be allocated to a species if the skeletal anatomy of that species is otherwise unknown. Some of the hominin

ABOVE The OH5 skull from Oldupai Gorge shows many characteristic features of *Paranthropus boisei* such as massive chewing teeth, flared cheek bones and a sagittal crest on top of the skull.

post-cranial fossils at Oldupai Gorge or Koobi Fora could potentially belong to either *P. boisei* or *Homo habilis*, as both species are known at these sites. In 2013 researchers reported the discovery of a partial skeleton from Oldupai Gorge (OH80), dating from just over 1.3 million years ago. The skeleton includes a handful of teeth, as well as bones from the arm and leg, so researchers could reliably identify it as belonging to *P. boisei*. The arm and leg bone reveal an extremely robust body form. This individual weighed approximately 46 kg (101 lb), about the same as the largest individuals from *P. robustus*, and is thought to have been a chunky male. One of the leg bones has pits on the surface that resemble tooth marks, suggesting that the hominin may have been eaten by carnivores. Comparison with the OH80 skeleton has enabled researchers to assign other isolated bones and partial skeletons to *P. boisei*. A recent find of associated arm and hand bones from Ilerat in Kenya confirmed that *P. boisei* had a strong upper limb together with a hand capable of making and using tools.

Living side by side

Paranthropus means 'alongside humans' and all of the species in this group lived alongside other species from *Australopithecus* or *Homo*. Various combinations of *Paranthropus* and *Homo* species appear to have coexisted at sites in both eastern Africa and South Africa. Reliable access to sufficient food at all times of year is the first prerequisite for survival so it is likely that these species were targeting different food resources. When two species that are similar in many aspects of their biology live side by side, as is the case with gorillas and chimpanzees in some African forests, they often have different preferred foods or fall back foods, allowing them to divide up the available resources. Species that share the same landscape in this way are termed 'sympatric' and the anatomical differences between them often reflect adaptations to different diets. *Paranthropus* may have specialized in chewing vast amounts of low-quality vegetation, possibly supplemented by termites, whereas *Homo* opted for high-quality foods that were easier to chew and digest but more difficult to procure.

The occurrence of fossils from more than one species at a single site does not necessarily mean that they were there at the same time. It does nevertheless make life more complicated for archaeologists, because stone tools or evidence of butchery could have been made by either species. And it complicates matters for paleoanthropologists, because isolated bones cannot always be allocated to a particular species. Associations between fossils of different species and other objects are also sometimes misleading, and scientists look carefully for clues about how an assemblage formed. For example, some accumulations of fossils and stone tools in open air sites arise due to the actions of flowing water. Water flow can transport objects far from their original location and bring together objects that were not originally associated with one another. Heavier material is left behind in areas where the water flows more slowly, and lighter bones are carried further. These processes are thought to have been acting at Oldupai Gorge. Predators and scavengers sometimes carry carcasses or bits of carcasses of hominins and other animals away from the places where they were killed. The leftovers from carnivore meals can be recognized from the characteristic marks on the bones produced by teeth, or because the profile of skeletal parts is what would be expected after a carnivore has eaten the softer parts. As we have seen, many hominins were killed or scavenged by carnivores and sometimes the evidence that they were dragged or snatched from the places where they were living is unambiguous.

CHAPTER 5

The origins of humans (genus *Homo*)

SCIENTISTS USE 'HUMAN' in different ways, but in this book we equate its use with membership of the genus *Homo* ('Man'). Defining what we mean by 'human' is not a straightforward task, however, as it goes beyond the structural features that are physically preserved in fossils. Humans today show great behavioural complexity compared with other animals, and this is especially apparent in our reliance on tool-making and tool-using in everyday life (of course most of us today don't actually make the tools we use, but rely on others to make them). Indeed, humans used to be defined as 'Man the Toolmaker', and the appearance of stone tools at the beginning of the archaeological record was viewed as fundamental evidence for the emergence of humanity and our genus *Homo*. However, we now know that birds like crows, as well as various primates, can make and use simple tools, so the distinction is not absolute. In addition, there is recently discovered evidence from Lomekwi in northern Kenya that large stone flakes of lava were being manufactured as far back as 3.3 million years ago (see p. 68), long before there is any accepted fossil evidence for the genus *Homo*. Such flakes may well have been made by an australopith, an informal name for species of hominins belonging to the genera *Australopithecus*, *Paranthropus* and their closest relatives.

OPPOSITE Geological deposits in the Hadar region of the Afar depression, north-eastern Ethiopia, span more than a million years of late Pliocene and early Pleistocene time. Although most famous for finds of *Australopithecus afarensis*, such as the Lucy skeleton, Hadar also produced upper jaw A.L. 666-1, dated to about 2.3 million years ago, which has been assigned to early *Homo*.

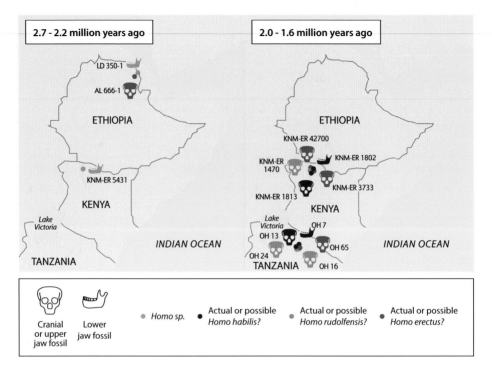

ABOVE This chart shows some of the evidence for distinct human and human-like species in East Africa about 2 million years ago. The classification of a number of these fossils is controversial.

Complex language has also been proposed as a defining feature of humans. While apes can learn to use large numbers of symbols to transmit basic information to humans or other apes, they cannot deal with abstract concepts, or talk about the distant past or future. How and when our ancestors progressed from ape-like language capabilities to those of humans today is very difficult to read from the archaeological record. It was surely there in the hunter-gatherers who produced sophisticated cave art in Europe 35,000 years ago, but estimating human language capabilities some 500,000 or a million years ago is almost impossible.

We are on somewhat safer ground when it comes to the bodily features that are preserved in fossils. Habitually walking on two legs is certainly a special human adaptation, but as we have seen, it had actually developed much earlier in human evolution. Our long-legged body shape, with a human-shaped rib cage and the loss of climbing adaptations in the upper body, is perhaps more promising as a defining 'human' feature, since it seems to have originated more recently, perhaps only about 2 million years ago. A large brain in relation to body size is another human characteristic, and this can be estimated very accurately in well-preserved fossil braincases. However, the old view that there was a 'cerebral Rubicon' that clearly separated all humans from all apes (or australopiths) has been repeatedly undermined by discoveries of small-brained fossil specimens that display many other human traits. Examples of this are *Homo erectus* (p. 94), *Homo naledi* (p. 101) and *Homo floresiensis* (p. 132).

Various other characteristics have been proposed as markers of the appearance of true humans. They include a reduction in the size of the teeth, jaws and associated musculature, the relative size and projection of the face, prominent nasal bones, and the evolution of the longer childhood growth period that characterizes humans today. In reality, many of these features could have evolved gradually, and at different rates, and thus would not have appeared suddenly as a 'package'. Moreover, if a trait such as longer legs had selective value 2 or 3 million years ago, it could have evolved in parallel in different lineages. In turn, this would make it more difficult for us to judge whether creatures with longer legs 2 million years ago inherited them from a common ancestor that had longer legs, suggesting the groups in question were closely related. Alternatively, if they had developed their longer legs independently of each other, they should not be classified together based on that trait. Problems like this make the recognition of the first 'humans' as problematic today as it was in 1964, when *Homo habilis* was named as the earliest human species.

Homo habilis and *Homo rudolfensis*

Simple stone tools made of lava had been found in the deepest deposits at Oldupai Gorge in Tanzania long before any signs of their possible manufacturer were discovered. In 1960, famous palaeoanthropologists Louis and Mary Leakey excavated hominin fossils that were clearly distinct from the *P. boisei* cranium that Mary had found a year earlier (see p. 78). These included a distorted mandible with teeth and parts of a braincase, as well as hand, foot and leg bones, representing both adult and young individuals. The front teeth were relatively larger and the back teeth relatively smaller than in australopiths, and thus more human-like. The skull bones were thin and large in area, suggesting a brain that was larger than in australopiths and modern apes. By 1964, Louis Leakey and his co-authors had decided that the fossils could justifiably be named as a new species of the genus *Homo*. They named the species *Homo habilis*, meaning 'Handy Man', because of an assumed association with the oldest Oldupai stone tools. Both these human fossils and the oldest artefacts from the Gorge are now dated to about 1.9 million years old.

The naming of this new and very primitive human species was highly controversial, but further discoveries from Oldupai and elsewhere gradually gave it greater scientific credibility. In the late 1960s Richard, one of the next generation of the Leakey family, initiated a research project at Koobi Fora, on the eastern side of Lake Turkana (formerly Lake Rudolf) in northern Kenya. He was soon rewarded with finds of stone tools like those recovered from the earliest layers at Oldupai Gorge (as well as remains attributed to the genus *Paranthropus*), dated at nearly 2 million years old. In 1972, the discovery of the fossil skull known as KNM-ER (Kenya National Museum-East Rudolf) 1470 seemed to further parallel the situation at Oldupai, since it was much larger brained (estimated volume about 750 cm^3) than any australopith, leading to suggestions that it, too, represented *H. habilis*. However, the tooth sockets suggested it must have had quite big teeth, and its face was large and very flat, with high cheek bones.

As the number of samples of early *Homo* fossils have increased at Koobi Fora and other East African localities, so has the variation they display in brain size, face size and shape, tooth size and jaw robusticity. This has led to suggestions that the larger hominin represented by the 1470 cranium should be referred to as a distinct species, *Homo rudolfensis*, while the smaller species would retain the original name given by Louis Leakey and colleagues, *H. habilis*. Because of this increased complexity, it is no longer clear which, if any, of these early forms of *Homo* might be ancestral to later humans, and which were the toolmakers.

Despite the hand, foot and leg bone fossils from Oldupai published in 1964 and attributed to *H. habilis*, it was still largely unclear what kind of skeletons these early human-like forms had. Some long leg bones were found in levels containing partial skulls and jaws of *H. rudolfensis* at East Turkana, suggesting human-like body sizes and proportions for this species. But it is equally possible that these and other isolated fossils belong to another species, *Homo erectus* (see p.94). Moreover, a subsequent discovery at Oldupai paints rather a different picture for *H. habilis*. OH (Oldupai Hominid) 62 consists of an upper jaw that has human-like tooth proportions, but the associated skeleton is very small and its build is decidedly non-human. The arms are relatively long compared with

THE ORIGINS OF HUMANS (GENUS *HOMO*) • 87

ABOVE Replicas of two crania from East Turkana, Kenya that are often attributed to *Homo rudolfensis* (KNM-ER 1470, top) and *Homo habilis* (KNM-ER 1813, bottom).

the leg bones, and this implies that the body proportions of this creature were ape-like, more like that of the Lucy skeleton from Ethiopia, which is over a million years older. The human status of *H. habilis* is therefore still controversial. Equally, the large face, large back teeth and thick jaws of *H. rudolfensis* do not look particularly human, and some scientists have made an evolutionary link with the 3.5-million-year-old species discussed earlier, *K. platyops* (see p.62), implying that these could represent a separate hominin lineage.

Despite lingering doubts about the human status of *H. rudolfensis* and *H. habilis*, some experts consider that true humans nevertheless go back even further than 2 million years. An upper jaw from Hadar in Ethiopia dated at about 2.3 million years old certainly shows human-like tooth sizes and proportions, while an even older find from the same region of Ethiopia – a partial lower jaw from Ledi-Geraru dated at about 2.8 million years old – is described as human-like in its teeth and its reduced size compared with autralopiths.

ABOVE Replica of articulated foot bones from Oldupai Gorge, assigned to *Homo habilis* (Oldupai Hominid 8).

THE IMPORTANCE OF TOOLS

ABOVE It's likely that some of the first stone tools were actually made and used by australopiths.

Because of the dominance of stone tools in the surviving record, archaeologists talk about the 'Stone Age', before the use of metals. The earliest period, the Old Stone Age, or 'Palaeolithic' (currently spanning from about 11,000 to 3.3 million years ago), is often divided into three stages, although these European-derived names are of limited value in describing the distinctive archaeological records of more distant regions. The Lower Palaeolithic (about 300,000–3.3 million years ago) is the earliest period of stone tool production, beginning with simple pebble and flake tools. About 1.7 million years ago, some sites in East Africa start to show evidence of large tools worked on both sides. We call these tools handaxes, or 'Acheulian', after the French site of Saint-Acheul, where they began to be recognized in large numbers from the 1850s onwards. Handaxes are usually almond or teardrop-shaped, and they were a very enduring artefact, made of local volcanic lavas in East Africa more than 1.6 million years ago, and native chert or flint rocks in Europe 300,000 years ago. We do not know all the functions for which handaxes were made, but associations of handaxes and large mammal skeletons show that they were certainly used for butchering carcasses.

The next stage, the Middle Palaeolithic (about 40,000–300,000 years ago), covers the stone tool industries made by the Neanderthals in Europe and western Asia, and it is sometimes also applied to those made by early modern humans in Africa and western Asia. The final stage of the Old Stone Age, the Upper Palaeolithic (about 11,000–42,000 years ago) is almost exclusively associated with modern humans, and is generally used for industries in western Eurasia. A comparable system of classification, based on modes of production, distinguishes early tool industries as Mode 1, handaxe industries as Mode 2, levallois production (see p. 90) as Mode 3, Upper Palaeolithic technology as Mode 4, and microliths (very small tools, often mounted on handles) as Mode 5.

ABOVE A refit of flint flakes from the site of Crayford, UK, about 200,000 years old.

THE IMPORTANCE OF TOOLS

However, it must be remembered that early humans did not always operate in accordance with the Palaeolithic and Mode stage names that we have created and use today. Thus, as part of the normal variation in human behaviour, artefacts that we consider characteristic of one period may also turn up in another. For example, Middle Palaeolithic/Mode 3 industries may not only have the expected stone tools, but also on occasion simple flakes or choppers like the earliest Palaeolithic Mode 1, handaxes characteristic of the Lower Palaeolithic Mode 2, blades supposedly typical of the Upper Palaeolithic Mode 4 and even, sometimes, Mode 5 microliths, which are supposed to occur in the much later Mesolithic period (Mode 5). Nor are artefact types necessarily correlated with particular kinds of humans. At least four different human species must have made and used handaxes, ranging from *Homo erectus* and *H. heidelbergensis* to both early Neanderthals and early *Homo sapiens*.

The types of stone tools discussed so far are generally grouped as Lower Palaeolithic/Modes 1 and 2, but from about 300,000 years ago, early Neanderthals in Eurasia and *Homo sapiens* in Africa invented a new way of making tools using a method we call levallois, after the Levallois-Perret suburb of Paris, where some of the earliest finds were recognized. Its appearance marks the arrival of the stage we call the Middle Palaeolithic/Mode 3. The technique allowed the tool-maker to map out the final shape of the required flake, which could then be struck off from its core with a single blow. This allowed much more control over tool production, and was probably also of benefit when groups were highly mobile, and needed to carry their raw materials across the landscape. The levallois method was probably the most important innovation of the Middle Palaeolithic, although to begin with, it was often used to produce traditional-looking handaxes.

Later, it was used in a variety of local stone industries in Europe, Asia and Africa.

In Europe, the Middle Palaeolithic industries of the Neanderthals are also known as 'Mousterian', after the French cave of Le Moustier, one of the first sites where they were recognized. The Neanderthals made different kinds of flake tools, which archaeologists call by names like 'knives', 'scrapers' and 'points', although we cannot be certain how the tools were used. Rarely, parts of wooden spears have been preserved at Neanderthal and pre-Neanderthal sites. For example, the end of a wooden spear was found within an elephant skeleton at Lehringen in Germany. The Neanderthals probably also mounted stone points on wooden handles, binding them with plant fibres or natural resins, to make short stabbing spears. They almost certainly made other kinds of artefacts from wood, and processed skins to make simple clothing. They seem to have made much less use of bone, antler and ivory as tools, however, even though the materials were all around them. It is likely that these materials were more difficult to work without the specialized stone tools that become more common in the subsequent Upper Palaeolithic.

Although imported natural pigments such as manganese (black) and iron (red) oxides are sometimes found in large quantities at their sites – suggesting these were used to colour objects, or their bodies – there is only limited evidence so far that the Neanderthals decorated the walls of their caves, and no firm evidence that they produced representational or figurative art. However, there are some examples of geometric patterns made by Neanderthals on bones and stones. In addition, certain butchery patterns found on bird bones from several southern European sites suggest the Neanderthals were specifically removing the feathers, perhaps for body adornment. Curiously, there is rare evidence of a walled structure of unknown purpose deep within Bruniquel Cave in France. This was made from broken stalagmites about 175,000 years ago, when only Neanderthals are known to have been living in the region. As the finds accumulate we are gaining a picture of a more sophisticated human than we first surmised. However, recent claims that a site in Slovenia contained part of a Neanderthal flute have been refuted by studies suggesting that the bear bone in question had been punctured by carnivore teeth rather than by a Neanderthal tool.

Despite claims to the contrary, it seems very likely that the Neanderthals buried their dead in caves where they had been living. At sites in Europe and western Asia, numerous Neanderthal skeletons have been excavated in circumstances suggesting intentional burial. Some burials are associated with animal bones or imported stone blocks, suggesting possible ritualistic behaviour, but claims that Shanidar Cave in Iraq was the site of a 'flower burial' are less likely. In this case it seems that gerbils may have transported flowerheads deep into their burrows as nesting materials! As yet, there is no definite evidence of burials by ancient humans in the Far East, but the record is poorer than further west.

In Africa, the Middle Palaeolithic is alternatively known as the 'Middle Stone Age', often abbreviated to MSA. There is significant evidence of the symbolic use of red ochre pigment and bone working in southern African MSA sites 75,000 years ago, presumably indicating increasing complexity in behaviour. However, some of the first early modern humans known to have emerged from Africa – those found at the Israeli sites of Skhul and Qafzeh, around 100,000 years ago – were fundamentally similar to the Neanderthals in their technology. Nevertheless, as mentioned elsewhere, there are hints of a greater behavioural complexity in the imported red pigments, jewellery made from pierced sea shells, and the burial patterns, including a man at Skhul buried with a large pig jaw in his arms, and a child at Qafzeh buried with an antlered deer's skull.

LEFT Several flutes have been discovered in German caves from early *Homo sapiens* occupations during the Aurignacian cultural phase of the Upper Palaeolithic, dating from about 35–40,000 years ago. Most of them were made from bird bones, but this example from Geißenklösterle Cave (two views of one flute) has been manufactured from a much more difficult material – mammoth ivory.

THE IMPORTANCE OF TOOLS

About 42,000 years ago, there was a change in the dominant method of tool-making in Africa and western Eurasia. Whereas the usual procedure in the Lower and Middle Palaeolithic was to reduce a piece of rock down to only one or a few tools, Upper Palaeolithic techniques allowed many long thin flakes (or 'blades') to be systematically produced from a single original block of stone. They were then worked further to turn them into specialized tools that archaeologists call 'chisels', 'borers' 'knives' 'scrapers' and so on. In Europe and western Asia these industries are termed Upper Palaeolithic, and in Africa, Later Stone Age.

Alongside the predominance of blades, there was also an associated increase in the working of antler, ivory and bone, and even some evidence of clay working, ropes and weaving. Composite tools made of several components become more common, such as harpoons with detachable heads, and spear-throwers or atlatls, which were used to increase the throwing range of projectiles. There is increased evidence of the use of pigments, sometimes painted on objects, sometimes on cave walls, and sometimes on bodies at burial. Some archaeologists regard this 'creative explosion' as marking the definitive arrival of fully modern minds, although not all parts of the world show the arrival of the full suite of Upper Palaeolithic/Later Stone Age features at the same time. For example, bone-working, pigment use, shell and bone jewellery and complex burials are known from Australia at more than 30,000 years ago, but blade tools of European and African type are missing.

By 30,000 years ago in Europe, campsites were generally growing in size and duration of occupation, with evidence of larger dwellings made of wood, animal hides and even mammoth bones, where wood was unavailable. Fire technology became more complex with the production of stone-lined hearths and ovens, and oil-lamps carved from rock have been found deep in some European caves. Techniques of food gathering also diversified, with the development of water craft and systematic fishing, probably accompanied by the manufacture of nets, traps and pits. There also seems to be evidence for the beginnings of social stratification, since some individuals were buried with richer accompanying grave goods than others. About 35,000 years ago, at Sunghir in Russia, the skeletons of two adults and two children were excavated along with thousands of ivory beads that must have decorated the clothes in which they were buried. Not only would these beads have taken many hours to produce, accompanying javelins carved from mammoth tusks probably required weeks of production. All of this suggests that these children were the offspring of important individuals within their society. Burial patterns were also increasingly complex such as that seen at Dolní Vestonice in the Czech Republic, where three male teenagers, including one who was disabled, were buried together. The burial pit and the bodies were carefully arranged, with the addition of red ochre pigment and objects such as wooden stakes.

From about 42,000 years ago, there was a succession of Upper Palaeolithic industries in Europe, most of which have been named after the French sites where they were first recognized. One of the earliest – the Aurignacian – occurred widely across the continent from about 40,000 years ago, and is associated with some of the first modern Europeans ('Cro-Magnons') and some of the earliest known representational art. In parts of Europe this was succeeded by the Gravettian, known from famous sites such as Predmostí and Dolní Vestonice, while industries such as the Solutrean and Magdalenian followed in turn (the famous

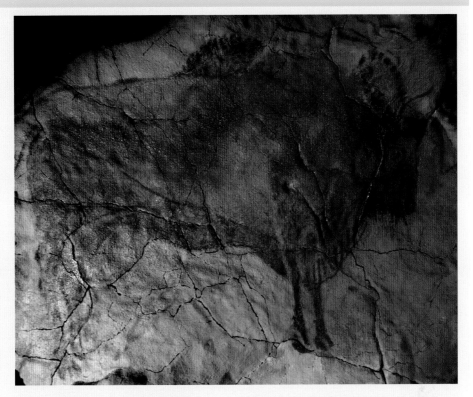

ABOVE This beautiful polychrome painting of a bison was one of the earliest recognised examples of Palaeolithic cave art. It is on the ceiling of Altamira Cave in northern Spain, and is thought to have been painted about 15,000 years ago.

caves of Lascaux and Altamira were painted by Magdalenian people). The Magdalenian continued until some 11,000 years ago, beyond the end of the last glacial stage, after which the Upper Palaeolithic gave way to the Mesolithic (or Middle Stone Age, not to be confused with the much earlier African Middle Stone Age).

What is particularly intriguing is that in France and Spain, there was a distinctive early Upper Palaeolithic industry, the Châtelperronian, dated to about 42,000 years ago. This has signs of significant pigment use and the production of body decorations made from pierced animal teeth. Several lines of evidence (human fossils and ancient DNA) strongly suggest that it was manufactured by Neanderthals. So, did the Neanderthals independently invent this version of the Upper Palaeolithic, did they 'borrow' innovations from neighbouring Cro-Magnons, or did they even develop more complex behaviour because they had received modern human DNA from interbreeding? The answer should become clearer as researchers continue to focus their efforts on this fascinating question.

Becoming human

So, recognizing where 'humans' began in the fossil record is no easier than it was when *H. habilis* was named and described in 1964. In fact the sheer diversity of fossils around 2 million years ago actually makes the situation even more complex. Are we defined by our small jaws and teeth, our large brain, our long legs, habitual tool-making and meat-eating, or some combination of these or other traits? If we require the combined presence of several traits to recognize a fossil as human, many of these early *Homo* specimens are simply too incomplete to make a confident diagnosis, and that is true overall until we arrive at the more complete remains and behavioural evidence of *Homo erectus*, which we discuss next. Is the diversity we see by 1.8 million years ago the result of evolution from a single primeval species of *Homo* more than 2 million years ago, or could the different '*Homo*' lineages have evolved in parallel from separate australopith-like ancestors? The human-like features of *Au. sediba* at around 1.95 million years ago, which we discussed earlier, are likely to have developed independently of the processes that produced early *Homo* in East Africa, showing that selection may have been driving the evolution of 'human' traits in different species at this time. If this is so, and some of the different *Homo* lineages had separate origins, they cannot all justifiably be assigned to the one genus *Homo*.

Homo erectus and the spread from Africa

The first discoveries of *Homo erectus* in the 1890s were stimulated by the views of German biologist Ernst Haeckel, who disagreed with Charles Darwin and Thomas Henry Huxley that Africa was the most likely cradle of humanity. Instead, Haeckel predicted that Southeast Asia would be a more promising location to look for the remains of the hypothetical creature he dubbed *Pithecanthropus* (Ape-Man). Eugene Dubois, a Dutch doctor, was a supporter of Haeckel and got himself posted to what was then the Dutch East Indies, to search for the evidence to prove Haeckel right. Within a couple of years, on the island of Java (Indonesia), he recovered fossil teeth, a fragmentary jawbone and most importantly, a skull-cap and a femur. While the skull-cap was long and low, with a strong brow ridge and a brain size well below that of most living *Homo sapiens*, the thigh bone had a bony outgrowth due to disease, but was human-like, indicating an erect posture. Hence, Dubois named his find

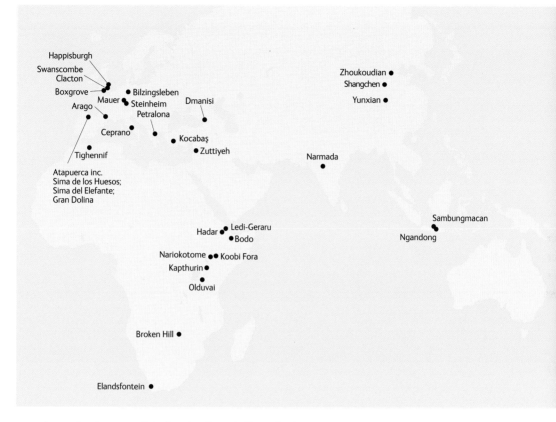

ABOVE A map showing some of the key sites for early *Homo* fossils and archaeology.

Pithecanthropus in honour of Haeckel, but added the species name *erectus*. We now recognize this species as part of the human genus and it has been renamed as *Homo erectus*. Although some researchers believe that what we now know as *erectus* consists of several distinct species, most accept a broad diagnosis of the species, which makes *erectus* the most long-lived human species at more than 1.5 million years, and the most geographically widespread until the dispersal of *H. sapiens* across the world.

Homo erectus individuals were already comparable to humans of today in body size and shape. The estimated adult weight ranged from about 41–65 kg (90–143 lb) while the height was about 1.4–1.8 m (4½–6 ft). Individuals had much wider hips than modern humans, a more voluminous rib cage, and greater muscularity. They had thicker skulls characterized by smaller brains on average (about 550–1250 cm^3 in volume), larger faces capped by a strong brow ridge over the eye sockets, and chinless lower jaws. The brow ridge is an especially intriguing structure, present in virtually all archaic humans, and reaching its maximum size in *H. erectus*. It has generated much debate about its possible function (if any), with most ideas concentrating on its role as a buttress or a means of dissipating forces passing through the skull. However, recent research suggests that it may instead have had a role in social signalling between archaic human individuals, enhancing messages sent by the eyes, whether friendly or aggressive.

Technologically, early *H. erectus* was equipped with a simple stone tool kit that looks best suited to scavenging meat, though later *H. erectus* peoples produced a wider range of stone tools and may well have been active hunters. Some experts believe that the more linear and long-legged physique of *H. erectus* evolved to facilitate long-distance movements (jogging and running) across open country to acquire meat. Nevertheless, it is likely that plant resources were also important to *H. erectus* throughout its timespan.

The earliest fossils that are complete enough to display the anatomical pattern of *H. erectus* are from East Africa and western Asia, dating from about 1.5 to 1.9 million years ago. The African fossils include several crania from East Turkana and one nearly complete skeleton of a boy aged about 9 years old, from Nariokotome in West Turkana. The conventional view is that the species evolved in Africa about 2 million years ago, either from a late form of *Australopithecus* or one of the more primitive forms of *Homo*. Soon after, it is thought to have dispersed to western Asia (as represented by fossils at the Georgian site of Dmanisi), and then to eastern Asia (including China) and Southeast Asia (including Java), although tools from Shangchen in China dated to about 2.1 million years old indicate that this could have happened even earlier. Those dispersals were probably enabled by environmental changes that gave early humans feeding opportunities over wider distances, and it is likely that their spread as far as Java, across what are now islands of Southeast Asia, was possible because of land bridges existing at that time.

We now know that *H. erectus* was a long-term inhabitant of the island of Java in Indonesia, with fossils dating from about 1.6 million years up to at least 250,000 years ago, while the species may have been present in parts of China until at least 300,000 years ago. The Chinese material includes a large collection of fossil braincases and fragmentary remains of the rest of the skeleton from the

ABOVE We don't know when humans were first able to make fire at will. As this reconstruction shows, it is likely that early humans first captured natural fires, and kept them alight for as long as they could.

site of Zhoukoudian, near Beijing, which were mainly excavated in the 1930s. These fossils are believed to show occupation at this site from at least 700,000–400,000 years ago. However, the precise date of the disappearance of the species in Indonesia and mainland China is currently unknown, and there are stone tools perhaps 150,000 years old on the island of Sulawesi, the makers of which are currently unknown. There is disputed evidence of the very late survival of *H. erectus* in Java in the form of fossil braincases and a few other fragments from the sites of Ngandong and Sambungmacan, estimated to be about 110,000 years old. These fossils show the characteristic *H. erectus* anatomy, but display brain sizes up to 50% larger than earlier examples of the species from the island.

From 1991, surprising new finds started to be made in Georgia (western Asia), at Dmanisi. Excavations under the remains of a medieval village yielded extinct

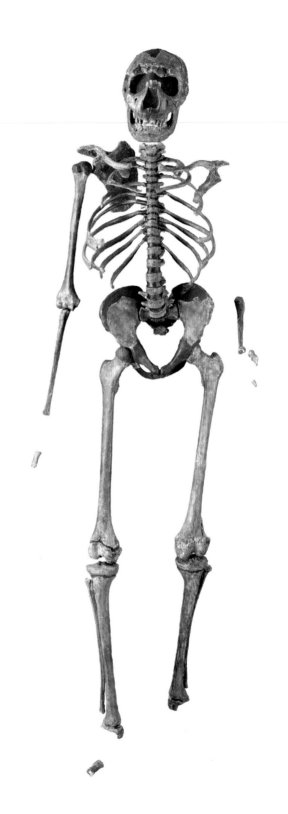

RIGHT This skeleton of a *Homo erectus* boy was discovered at Nariokotome, West Turkana, in 1984. Study of the KNM-WT 15000 skeleton shows that although Turkana Boy was only about 9 years old, he was nearly fully grown.

fauna including ostriches, primitive deer, rhinoceroses, large carnivores and a human lower jaw. Further excavations and research have placed the material at about 1.8–1.85 million years old, and the site has now produced five small-brained human crania, four more jawbones and many other parts of the skeleton, as well as simple stone tools. The morphology of the skulls indicates a very primitive version of the *Homo erectus* morphology, while the leg bones are relatively small, indicating an adult body size of only about 1.5 m (5 ft) and a weight of about 45 kg (99 lb). A few fauna had apparently also dispersed from Africa, including two species of sabre-toothed cats. These specialized carnivores lacked the teeth to strip a carcass clean of its meat, or break the thicker bones of their prey, so might have provided scavenging opportunities for early humans following them out of Africa. However, wider faunal comparisons suggest that the Dmanisi assemblages most closely resemble the contemporary forest and grassland fauna

ABOVE Excavations in the 1920s and 1930s at the Zhoukoudian Caves near Beijing, northern China, produced a large sample of *Homo erectus* fossils, dating from around 700–400,000 years ago. Unfortunately many of those fossils were lost during World War II, but Chinese excavators have recovered further finds at the sites since 1949.

LEFT This beautiful fossil, Skull 5 from Dmanisi, shows a probable male *Homo erectus*, with large jaws and a very small brain.

of southern Europe, supporting at least the possibility of an early extension of human settlement to that region.

Some researchers argue that the distinctive features of the Dmanisi fossils (including the smaller body and brain sizes ranging from only about 546–730 cm^3) indicate that this is a more primitive species than *H. erectus*, which could be called '*Homo georgicus*'. However, others feel that these features merely reflect its position as a very early member of the species *H. erectus*. Furthermore, some experts have argued that the primitive nature of the Dmanisi fossils and their great age may indicate that *H. erectus* originated in western Asia from a currently unknown ancestral species, and then migrated both eastwards to China and Java, and southwards into Africa. However, others point out that there are some fragmentary skull, pelvis and leg bone fossils from East Turkana that are dated to about 1.9 million years, as well as a child's cranium about 2 million years old from Drimolen in South Africa and the previously mentioned upper jaw from Hadar dating to about 2.3 million years, which could represent even earlier forms of *H. erectus*. Yet another taxonomic scheme proposes that the early *H. erectus* specimens from Africa and Georgia are sufficiently distinct in their primitive features (such as their teeth, their smaller overall size, and their thinner, less robust cranial walls) to warrant classification as a distinct species called *Homo ergaster* (Work Man).

Homo naledi

In 2015, a large number of fossils from a cave system in South Africa were named as a new human species, *Homo naledi*, which displayed a unique combination of human and non-human traits throughout the skeleton. Two years earlier, the first finds had been discovered by cavers in an almost inaccessible chamber deep within the Rising Star cave system, about 40 km (25 miles) from Johannesburg. Following this, a team of excavators (all women) slender enough to access the site was assembled by palaeoanthropologist Lee Berger, and they proceeded to recover more than 1,500 human fossils. The specimens represented at least 15 individuals, ranging from infants to the aged. Moreover, another 130 fossils of the same species were subsequently discovered in a second chamber more than 100 m (328 ft) away in the cave system.

The material assigned to *H. naledi* (naledi means 'star' in the Sotho language) is enigmatic as its geological age has recently been determined as surprisingly young (about 300,000 years old), while the means by which these thousands of remains arrived deep in the cave system, well beyond any natural light, are currently the source of much debate. The collection displays primitive characteristics resembling australopiths and some fossils of *H. habilis*, such as the small brain size (460–610 cm^3), curved fingers, and the shape of the shoulder, trunk and hip joint. Yet the wrists, hands, legs and feet look more like those of Neanderthals and modern humans. Adult height is estimated at about 1.46 m (5 ft) and body-weight was in the range 39–55 kg (86–121 lb). While the teeth have some primitive features (such as increasing in size towards the back of the tooth row), they are also relatively small and simple, and set in lightly built jawbones. In several ways, the material looks quite like the small-bodied examples of *H. erectus* found at Dmanisi in Georgia, and some experts have argued that *H. naledi* actually represents this species rather than a new one. However, the rich *H. naledi* sample does include some distinctive anatomical details in parts like the thigh and thumb. Some of the *H. naledi* bones are also poorly known in

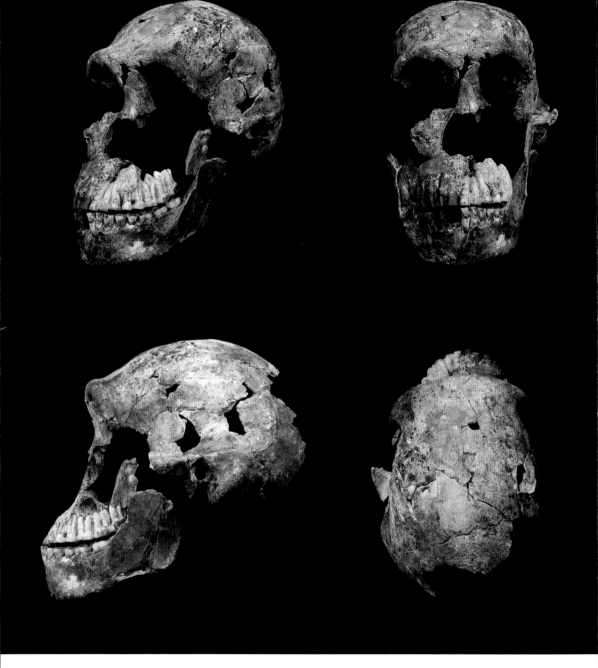

ABOVE This is the most complete skull and lower jaw of *Homo naledi* discovered so far. It displays the small brain and projecting face typical of this South African species.

other early human species such as *H. habilis* and *H. erectus*, so it is currently difficult to assess how similar these species were throughout their skeletons.

Despite the young date for the *H. naledi* fossils, their anatomy suggests that in evolutionary terms they could lie close to the origin of the genus *Homo* and of *H. erectus* itself, suggesting that this is a relic species, retaining many primitive

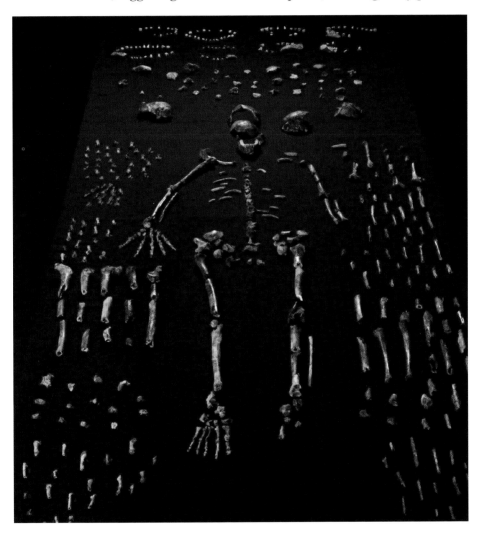

ABOVE An assemblage of most of the *Homo naledi* fossils from Rising Star Cave. About 15 individuals of adults and children are represented in this collection.

traits. Because *H. naledi* is currently only known from four sites within a single cave system, it is unclear whether it was restricted to southern Africa or more widespread. If it was more widespread, scientists might need to re-examine other diminutive fossils from across Africa that have often (and perhaps wrongly) been attributed to a small form of *H. erectus*. Additionally, although the *H. naledi* fossils are not so far associated with stone tools, it seems highly likely that its handiwork is present in the archaeological record of southern Africa, but currently unattributed.

The Dinaledi Chamber ('Chamber of Stars' in Sotho) is a cavity some 80 m (262½ ft) into the Rising Star system, and must have always been in darkness. The circumstances closely parallel those of a famous accumulation of around 6,500 human fossils found in the Sima de los Huesos ('Pit of the Bones') in the Sierra de Atapuerca in Spain (see p. 115). In both cases, there is no associated evidence suggesting the humans ever lived so deep in these caves. Because the

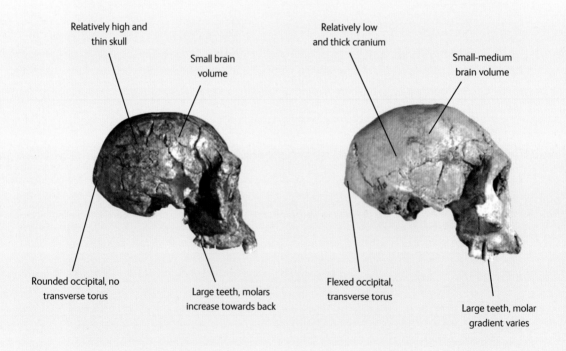

Atapuerca team recovered numerous bones of the hands, feet and spine that could be articulated – connected back together as they were in life – they proposed that the remains of at least 28 early Neanderthals had been intentionally thrown down into the pit, where the bodies had decayed. After considering alternative explanations, including whether the remains could have washed there or been dragged there by predators, Berger's team favoured a similar scenario for Rising Star to that proposed for the Sima accumulation. However, intentional and

BELOW A comparison of some typical features in the cranium and teeth of four *Homo* species (all replicas, except for *Homo naledi*, where a composite reconstruction has been made). The occipital bone at the back of the cranium shows variation in shape, sometimes being more smoothly rounded, in other cases being more sharply angled, with a ridge of bone (torus) across it.

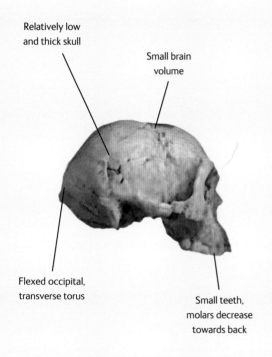

Relatively low and thick skull

Small brain volume

Flexed occipital, transverse torus

Small teeth, molars decrease towards back

Homo floresiensis

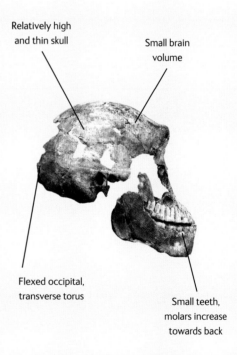

Relatively high and thin skull

Small brain volume

Flexed occipital, transverse torus

Small teeth, molars increase towards back

Homo naledi

repeated disposals of dead bodies deep within a pitch-black and dangerous cave is surprisingly complex behaviour for a creature with a gorilla-sized brain, so serious doubts still surround this interpretation. The mystery of the hominin deposition in the Dinaledi chamber is compounded by the recovery of remains of at least three more individuals in the Lesedi chamber, equally remote and dark, and over 100 m (328 ft) away.

Homo antecessor

Southern Europe has produced some fragmentary human fossils of more than a million years old. Two of them (the front of a jawbone and a finger bone) are from the Sima del Elefante site in the Atapuerca mountains of northern Spain, dated to about 1.2 million years ago. More than a century ago, a mining company cut a railway trench through the mountains there and in doing so, inadvertently exposed several ancient cave chambers that we now know are rich in fossil and archaeological remains. The human species represented in the Sima del Elefante is currently uncertain. It may be *H. erectus*, or a descendant form.

However, a distinct species – *Homo antecessor* (Pioneer Man) – is recognized from the neighbouring site of Gran Dolina. Simple stone tools and human material began to be excavated from this site in 1994, and palaeoanthropologists have now extracted the remains of several individuals, mainly children, including a partial skull, jaw fragments and teeth, and over 100 other parts of the skeleton. Because these fossils lie beneath deposits that preserve evidence of the last significant reversal of the Earth's magnetic field, they can be dated using the palaeomagnetism technique (see p. 33), to about 850,000 years ago. Most of the bones display butchery marks and damage consistent with them having been processed to extract meat and marrow, although the exact circumstances of this apparent cannibalism are, of course, unknown. The largely Spanish team who named and studied the material have suggested that *H. antecessor* is clearly distinct from *H. erectus* in features of the skull, jaws and teeth. The species does retain primitive features in the jaws and crowns and roots of the teeth, but some traits of the skull, collar bone and arm and leg bones are

OPPOSITE The Gran Dolina site at Atapuerca, northern Spain at an early stage of excavation. This is the only locality so far known to have produced fossils of *Homo antecessor*.

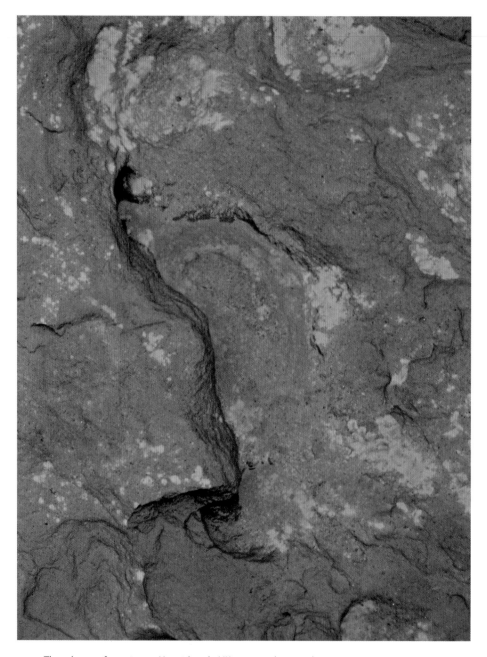

ABOVE These human footprints at Happisburgh, UK were made in mud along the banks of the ancient River Thames, about 900,000 years ago.

more typical of modern humans or Neanderthals. Even more controversially, it has been suggested that *H. antecessor* could have been the last common ancestor of the subsequent Neanderthal and *H. sapiens* lineages, with the flat face and hollowed cheek bones of several individuals presaging the morphology found in modern humans. This last idea has been challenged because *H. antecessor* seemed to be too ancient to represent the last common ancestor, which was believed to have lived only about 500,000 years ago. But recent research, discussed below, suggests that the last common ancestor of Neanderthals and modern humans may have lived well before 500,000 years ago, and additionally, microscopic study of the face in *H. antecessor* does indicate a resemblance to a modern pattern of growth, giving rise to a flattened face and delicate cheek bones.

The species is currently only known from the one site of Gran Dolina and there is a dearth of human fossils of comparable age from nearby regions. Three partial lower jaws and part of a cranium dating to about 700,000 years ago were found at Tighennif in Algeria in 1954–1955 (sometimes attributed to '*Homo mauritanicus*'), and a partial cranium attributed to *H. erectus* and dating to about 1.2 million years ago was recovered from a travertine quarry at Kocabaş in southwest Turkey in 2002. However, none of these fossils can be confidently linked with *H. antecessor* at present. Also, a set of human footprints dating to about 900,000 years ago were found in ancient river muds at Happisburgh in eastern England in 2013, but any link with *H. antecessor* can only be provisional unless actual fossil evidence is recovered.

Homo heidelbergensis

This species was named in 1908 from a chinless fossil jawbone found a year earlier at a sand quarry in Mauer, near Heidelberg in Germany. It is believed to date from about 600,000 years ago. Fourteen years after this find, a nearly complete cranium and a shin bone were recovered from the Broken Hill metal ore mine in what was then Northern Rhodesia (now Zambia). At first that cranium was designated as belonging to another new human species, *Homo rhodesiensis*, but in the last 30 years, both the Mauer jaw and the Broken Hill cranium have often been placed together in an extended diagnosis of the species *Homo heidelbergensis*. In the view of some scientists this species was widespread in Eurasia and Africa from about 300–600,000 years ago, with representative remains including fossils from Germany (Mauer and Bilzingsleben), England

(Boxgrove), France (Arago), Italy (Ceprano), Serbia (Mala Balanica), Greece (Petralona), Ethiopia (Bodo), Kenya (Kapthurin), Zambia (Broken Hill) and South Africa (Elandsfontein). It may also have existed in Asia, as indicated by fossils like Narmada (India) and Yunxian (China), but this is still unclear.

One of the major difficulties in uniting all this material under the name *H. heidelbergensis* is that most of the finds are of crania with or without faces and generally without lower jaws, while the type specimen from Germany is an isolated lower jaw, with some idiosyncratic features. Some scientists therefore argue that the *heidelbergensis* name is better restricted to the Mauer jawbone, while the larger group of fossils could be referred to *H. rhodesiensis*, based on the Broken Hill cranium from Kabwe, Zambia.

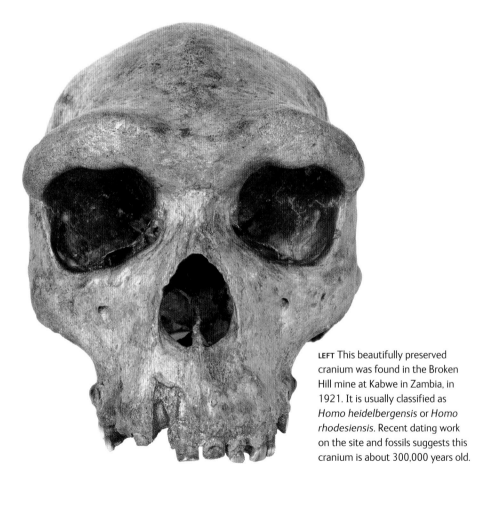

LEFT This beautifully preserved cranium was found in the Broken Hill mine at Kabwe in Zambia, in 1921. It is usually classified as *Homo heidelbergensis* or *Homo rhodesiensis*. Recent dating work on the site and fossils suggests this cranium is about 300,000 years old.

ABOVE Excavations during the 1990s at the Boxgrove site in southern England produced hundreds of handaxe tools and human fossils attributed to the species *Homo heidelbergensis*.

The site of Boxgrove in southern England is probably one of the most informative sites about the behaviour of humans (possibly the species *Homo heidlebergensis*) about 500,000 years ago. Boxgrove was already known for its evidence of handaxes and large mammal bones bearing butchery marks, but its importance was accentuated by the 1993 discovery of a human shin bone, and the subsequent find of two incisor teeth. Dating from about 480,000 years ago, these finds represent the oldest physical evidence of humans known from the British Isles, and have been allocated to *H. heidelbergensis*, though this assignment can only be tentative for such fragmentary remains. The shin bone is massively built, but the incisors (from a different individual) are more modest in size. Heavily worn, with a mass of scratches and pits on their front surfaces, they suggest that plant or meat materials were clenched in the jaws and processed by flint tools. The direction of the scratches suggests that the tools were being used by a right-handed person.

ABOVE A reconstruction of the butchery of a rhinoceros carcass at Boxgrove about 500,000 years ago. It is not certain that wooden spears were in use at this time.

LEFT This partial human shin bone from Boxgrove, UK had been chewed by a carnivore such as a wolf. It was from a large and heavily built individual, probably an adult male.

Homo heidelbergensis was typified by a large body size in males or assumed males. The estimated height and weight for the Boxgrove and Broken Hill individuals are about 1.75 m and 1.80 m (5¾ ft and 6 ft), and 76 kg and 72 kg (167½ lb and 158¾ lb), respectively. The brain size ranged from about 800–1,300 cm^3 and thus overlapped that of humans today (typically 1,100–1,500 cm^3), while the body shape was human-like, but with a very broad pelvis and ribcage, and probably also wide shoulders. The skull was large but long and low, with heavy brow ridges and a chinless lower jaw. It has been argued for many years that the species *H. heidelbergensis* was the probable last common ancestor of Neanderthals and modern humans, with the divergence of the respective descendant lineages in Eurasia and Africa developing after about 500,000 years ago. However, new fossil and genetic data could require a reconsideration of this view, as the common ancestor may have lived at an earlier date, and with a more modern-looking facial structure than *H. heidelbergensis* (perhaps more like that of *H. antecessor*, as discussed above).

The Neanderthals

The Neanderthals are the best known of all ancient humans, partly because of the large number of their sites that have been excavated, providing many thousands of their artefacts and fossils, including several nearly complete skeletons. We also know about their genetic make-up, as several Neanderthal genomes have now been reconstructed from their fossils. Although Neanderthal remains had been found previously at sites in Belgium and Gibraltar, it was the partial skeleton of a male Neanderthal rescued during quarrying operations in the Neander Valley in Germany in 1856 that was recognized as a distinct form of human, and named as a new human species, *Homo neanderthalensis*, 8 years later. We now know that the species ranged widely in Eurasia, from Portugal and Wales in the west across to the Altai Mountains of Siberia in the east. Neanderthal populations were also adaptable, living in cold steppe environments in England and Siberia 60,000 years ago, and in warm

RIGHT This reconstruction of a Neanderthal woman shows the short and broad body shape characteristic of the species.

temperate woodlands in Spain and Italy 120,000 years ago. Although the typical image of Neanderthals is of highly carnivorous ice-age hunters and scavengers, food remains preserved in the calculus (hardened tartar) around their teeth show that they also had a varied diet of plant foods, either collected directly or by eating the stomach contents of their plant-eating prey. In Gibraltar, they also sourced and consumed mussels and young seals and perhaps also dolphin, though that may well have been from scavenged carcasses.

The Neanderthals evolved over a period of at least 400,000 years, so in the following sections we will first discuss the earliest members of their lineage known from Europe, followed by the best-known Neanderthals, who lived between about 40,000–100,000 years ago.

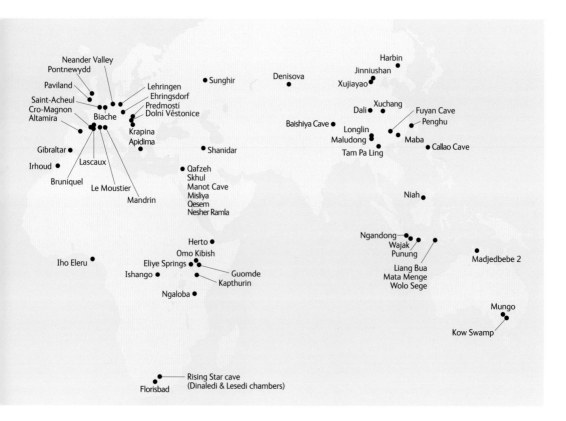

ABOVE A map showing some of the key sites for later *Homo* fossils and archaeology.

Early Neanderthals

The hills of Atapuerca in northern Spain are not only famous for the Gran Dolina site, which produced the *H. antecessor* fossils. Deep within them is the Cueva Mayor cave system and a chamber called, with good reason, the Sima de los Huesos ('Pit of the Bones'). Since 1976 many cave bear and human fossils have been recovered from the Sima, with the human material now numbered at over 6,500 fossils, representing about 29 individuals. The human remains consist of jumbled partial or nearly complete skeletons, mainly those of adolescents and young adults, and their mode of deposition in the Sima is still uncertain. There

LEFT The Sima de los Huesos site deep within a cave at Atapuerca, northern Spain has produced more than 6,500 human fossils, dating from about 430,000 years ago.

is no evidence the Sima was ever part of a living site, and there is only a single artefact, a handaxe made of non-local rock. Recently it has been recognized that some of the Sima skulls show signs of head injuries made around the time of death, so it seems possible that at least some of these individuals were killed and thrown into the pit via a now-inaccessible entrance to the cave system.

The Sima skeletons were previously claimed to be about 600,000 years old, and to represent *H. heidelbergensis*. However, they display clear affinities to subsequent Neanderthals in details of the skull, face, jaws and especially their teeth and are now dated to about 430,000 years instead. Moreover, these remains have yielded ancient DNA (deoxyribonucleic acid), which places them firmly on the Neanderthal genetic lineage, in line with their morphology. Their status as very early Neanderthals suggests that known *H. heidelbergensis* fossils might be

ABOVE The cul-de-sac of the Sima de los Huesos site contains many fossils of early Neanderthals and cave bears, and these fossils have yielded DNA, some of the most ancient yet recovered.

OPPOSITE This handaxe, nicknamed Excalibur, is the only artefact recovered so far from the Sima de los Huesos site. Some of the Atapuerca team believe that it was left as a tribute to the individuals deposited in the Sima.

too close in age to the Sima people to represent their immediate ancestor, since the descendant population (Sima) already displays a lot of Neanderthal traits. Additionally, some new genetic calibrations, partly based on the Sima data, place the Neanderthal-*sapiens* divergence at more than 600,000 years ago, closer to the time of *H. antecessor*. So, as mentioned already, it is currently uncertain whether the last common ancestor of the later species was more like *H. heidelbergensis* or *H. antecessor*, and on which continent that ultimate ancestor lived.

These finds not only provide a unique view of human biological variation about 400,000 years ago, but also flesh out the more limited fossil data from the rest of Europe. The adult cranial remains range in brain volume from about 1,100 cm^3 to 1,400 cm^3, and thus brain size certainly overlapped the Neanderthal and modern ranges. Asymmetry in the brain cavity and the arms compared with the patterns of scratch-marks on the teeth suggest more equal proportions of right- and left-handedness than in people today, where right-handedness predominates. The inner and middle ear bones are more comparable with those of modern humans, lacking some of the shape distinctions found in *H. neanderthalensis*, while the preserved hyoid bones from the throat, like that of the Kebara Neanderthal skeleton discovered in Israel in 1982, show no significant differences from those of recent humans, consistent with vocalizing abilities like people today.

The skull remains are comparable in overall shape with other similarly aged European crania such as those from Swanscombe in England and Steinheim in Germany and, as in the latter two specimens, at the back of the skull there is a small pit in the middle of the occipital bone called the suprainiac fossa, found in all Neanderthal fossils. The Sima crania are long, low and relatively large and broad facially, but they show a range in facial shape. Some are rather flat, while others show strong mid-facial and nasal projection, and inflated, retreating cheek bones, like those of Neanderthals. Moreover, microscopic studies of facial growth in young individuals from the Sima sample indicate a developmental pattern like that of Neanderthals. Mandibular and dental features combine some traits found in *H. antecessor* and *H. heidelbergensis* fossils with many more that can be found in the Neanderthals. Individuals were strongly built, particularly in the lower limbs, with wide hips and shoulders. Estimated height and weight of male and female individuals ranges between at least 1.60–1.70 m (5¼–5½ ft), and 52–90 kg (116¾–198½ lb), making them similarly sized to the later Neanderthals. The extensive dental remains are sometimes quite heavily worn, and large at the front, but not big by Neanderthal standards. Wear and scratch-marks on the incisors show that they clenched these teeth to hold and cut or manipulate meat or vegetable matter, as at Boxgrove. Interestingly, the back teeth also show signs of the use of toothpicks, presumably made of wood or bone.

ABOVE This beautifully made flint handaxe from Swanscombe, southern England, was found in the same levels as the Swanscombe 'skull'.

As already mentioned, the Swanscombe 'skull' (in fact the back half of a braincase, with a brain volume of about 1,275 cm^3) from the Thames valley in England is also generally regarded as an early member of the Neanderthal lineage. It dates from a warm interglacial period about 400,000 years ago, and the gravels at Swanscombe contain many thousands of handaxe artefacts made from flint. A slightly older site at Clacton in eastern England has also yielded the rare preservation of the tip of a spear carved from yew wood.

ABOVE The three skull bones found at Swanscombe between 1935 and 1955. They fit together perfectly, forming the back of a skull. Given their excellent preservation, it seems likely that more of the skull was present in the gravels, but not recovered.

ANCIENT CLIMATES

The Earth's climate has undergone numerous fluctuations in the last 4,000 million years, many of them quite extreme. These changes have continued during the last few million years that the human lineage has been evolving, and modern humans have been particularly fortunate to have enjoyed the last 11,000 years of relative climatic stability in the present interglacial period known as the Holocene. What is popularly known as the 'ice age' began more than 2.5 million years ago, but actually consisted of many cold stages (glacials) separated by warm stages (interglacials). These fluctuations can be recognized by chemical changes recorded in cores taken from the floors of the oceans, known as Marine Isotope Stages. The main driving force for the Earth's big swings in climate was deciphered by several scientists, including the mathematician Milutin Milankovitch, who realized that there were three main factors behind the growth and decay of the Earth's ice-caps: the degree of circularity of the Earth's orbit around the Sun; the changing tilt of the Earth's axis of rotation; and the varying time of the year when the Earth is closest to the Sun. These three drivers cycle about every 100,000, 40,000 and 20,000 years, and working in line or in conflict, they produce the large oscillations in our recent climates. When the ice-caps were at their largest, sea levels dropped globally about 125 m (410 ft), due to the amount of water locked up in these ice-sheets. At those times, Britain was physically connected to the rest of Europe, while New Guinea, Australia and Tasmania were united to form a single super-continent. These periodic changes in the Earth's circulating water system also affected the tropics and sub-tropics, with fluctuations in monsoon rains, and in forests, grasslands and deserts.

However, within the large-scale Milankovitch cycles there were also rapid and repeated 'millennial-scale' fluctuations lasting only centuries or a few thousand years. The causes of these are much less understood, but they may be triggered by factors like instabilities in the ice-caps and shorter term alterations in patterns of atmospheric and ocean circulation. For example, the Gulf Stream that flows from the Gulf of Mexico to northwestern Europe gives Britain its relatively mild winters, but this sweep of warm surface water has suddenly switched off several times in the past 100,000 years, chilling the North Atlantic and plunging the region into bitter cold. At times, it has also restarted its flow just as rapidly.

BELOW In cold stages during the last 300,000 years, northern Eurasia was covered by 'mammoth steppe', a productive ecozone that supported a rich fauna, including mammoth, horse and glutton.

ANCIENT CLIMATES

A timeline of key sites and events in the last 1 million years. The climate curve is a combination of deep-sea records, and corresponds to global ice volume. This gives an indication of changing climatic conditions through time.

Years before present ('000s)

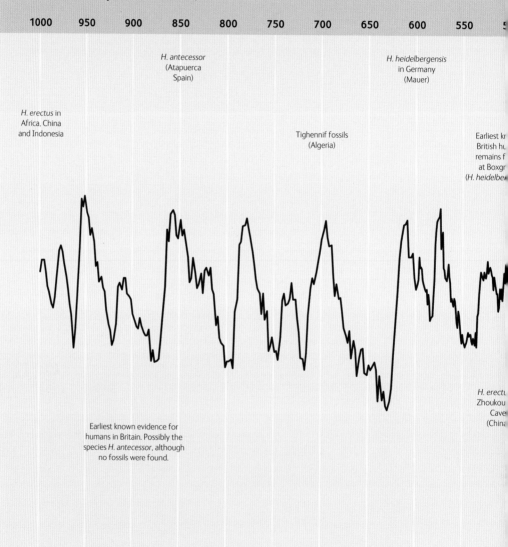

1000　950　900　850　800　750　700　650　600　550

H. erectus in Africa, China and Indonesia

H. antecessor (Atapuerca Spain)

H. heidelbergensis in Germany (Mauer)

Tighennif fossils (Algeria)

Earliest kn British hu remains f at Boxgr (*H. heidelbe*

Earliest known evidence for humans in Britain. Possibly the species *H. antecessor*, although no fossils were found.

H. erectu Zhoukou Cave (China

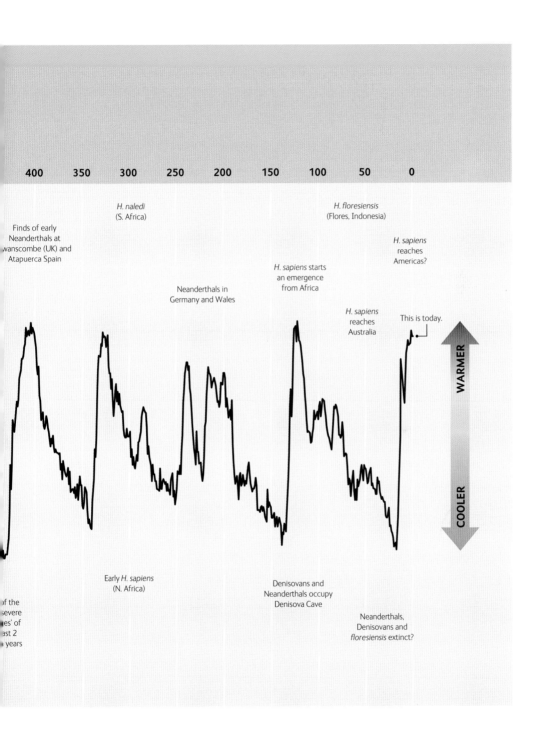

The late Neanderthals

Judging from the Sima and Swanscombe evidence, the Neanderthal lineage was already well-established in Europe by 400,000 years ago, and we can track the evolution of this group through subsequent fossils such as Krapina (Croatia), Steinheim and Ehringsdorf (Germany), Saccopastore (Italy), Pontnewydd (Wales) and Biache (France), through to the best-known examples from Europe and western Asia in the period 40,000–130,000 years ago. Fossils from the Israeli sites of Zuttiyeh, Qesem and Nesher Ramla may similarly track Neanderthal evolution in that region. The late Neanderthals retained the wide bodies of *H. heidelbergensis* and early Neanderthals, but were sometimes shorter in stature; the estimated adult height and weight ranged between about 1.50–1.75 m (5–5¾ ft) and 64–82 kg (141–180¾ lb), respectively. The wide and bulky trunk,

BELOW Excavations at Pontnewydd Cave in northern Wales spanning more than 100 years have produced 20 fossils of early Neanderthals, dating from about 230,000 years ago.

BELOW These artefacts from Pontnewydd, Wales, were made from local volcanic rocks by early Neanderthals.

and short bones of the lower parts of the arms and legs, gave the Neanderthals proportions that would have minimized the surface area of their skin, presumably to conserve heat under the predominantly colder conditions of the last 200,000 years. However, some researchers argue that this physique also gave the Neanderthals greater power in their arms and legs for close range ambushes of their prey during hunting, consistent with their larger ribcages, suggesting more voluminous lungs.

Their skulls were long and low with a definite brow ridge, but their brain volumes ranged from at least 1,200 cm^3 to 1,750 cm^3, larger than the modern average although not larger when body size is considered. (It's interesting to note at this point that *H. sapiens* skulls from 20,000 or 30,000 years ago also had larger brains on average than people today – the fact is that our brains have actually got smaller in the last 10,000 years!) The Neanderthal face was dominated by a wide and projecting nose, which some experts believe acted to moisten and warm the air they were inhaling. Their front teeth were somewhat enlarged, perhaps because they were regularly used like a third hand in the processing of food and other materials, and the teeth were placed forwards in the jaws, which generally

ABOVE A Neanderthal group camped at a rock shelter, engaged in butchery, tool-making, hide-working, wood-gathering and socialising.

lacked any sign of a projecting chin. Ancient DNA began to be recovered from Neanderthal fossils in 1997 (in fact the very first was obtained from the Neander Valley skeleton), and this has led on to the reconstruction of several complete genomes in the last few years. These indicate that Neanderthals ranging from Spain to Siberia were relatively low in numbers and diversity during their last 50,000 years, before they went physically extinct about 39,000 years ago. The genome of one female individual from the Altai also shows signs of long-term inbreeding in her population, a further indication of low numbers and isolation. This may be because regular and sometimes extreme climatic fluctuations (millennial-scale oscillations) continually fragmented Neanderthal groups, preventing them from building up large populations and continuous distributions across their range.

The Denisovans

Denisova Cave in the Altai region of Siberia has been under archaeological excavation for many years, but it received wider attention in 2010 with the recovery of a very distinctive mitochondrial DNA (mtDNA) sequence (see p. 8) from a human finger bone excavated there. The results suggested that this 'Denisovan' had diverged from the lineages of modern humans and Neanderthals nearly a million years ago. Soon afterwards, a genome was reconstructed using DNA from the same finger bone, which instead suggested that this Denisovan was more closely related to the Neanderthals. Further genetic sequences have now been recovered from three large-sized adult and immature molar teeth from different parts of the cave sequences, as well as from bone fragments and even the cave sediments. These showed that the Denisovans probably occupied

ABOVE Denisova Cave in Siberia had been under archaeological investigation for many years before fossil and DNA evidence of the Denisovans was uncovered. At times, this cave would have overlooked rich hunting territories in the valley below.

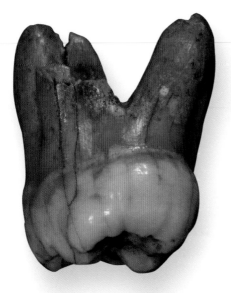

LEFT One of the large molar teeth discovered at Denisova Cave in Siberia. This is one of the fossils that produced DNA evidence of the distinctive Denisovans, who lived at the same time as Neanderthals and *Homo sapiens*.

BELOW The Maba fossil from southern China is about 300,000 years old. Although incomplete, it shows some shape resemblances to Neanderthal skulls.

the cave from about 50,000 to 200,000 years ago, with archaeological and DNA evidence suggesting that modern humans were at the site by about 45,000 years ago. However, untangling who occupied the cave in earlier periods, and made the Middle Palaeolithic stone tools found there, is complicated by the fact that other fragmentary fossils from the site, as well as the cave sediments have been shown to have Neanderthal, not Denisovan, genomes. Thus, there may well have been alternations or gaps in the Denisovan and Neanderthal presence in the cave over many millennia, followed by modern human occupation. New evidence reveals that Neanderthals and Denisovans certainly met in the vicinity since genetic data show a hybrid girl nicknamed 'Denny' lived there over 100,000 years ago.

So who were the Denisovans, and how do they relate to us and to the Neanderthals? Genetic data provide an interesting additional fact in that the Denisovan genomes not only show signs of past interbreeding with Neanderthals, but also with another, more archaic, species, which might explain both their large and primitive-looking teeth and the distinctive mtDNA sequences. The estimated divergence date from the Neanderthals is about 400,000 years, around the time of the Sima early Neanderthals.

Homo daliensis and other archaic Chinese fossils

Fossil fragments from Denisova Cave are all that are definitely identifiable as Denisovan, but it is possible that some Chinese fossils already known from between 100,000 and 350,000 years ago could represent Denisovans, including the skull remains from Dali, Maba, Xujiayao, Hualongdong and Xuchang, and the skull and large-bodied skeleton of a woman from Jinniushan. These fossils have proved difficult to classify, in some ways resembling *H. erectus*, in other ways *H. heidelbergensis* or the Neanderthals. A small but very robust lower jaw dredged from the sea near Penghu, Taiwan also resembled Denisovans in having large molars, and there was speculation that this, too, might represent a Denisovan. These suspicions have been supported by the discovery of a similar jawbone at least 160,000 years old from China, this time from Xiahe, on the Tibetan plateau. Although no ancient DNA could be recovered from it, the jawbone was analyzed for its fossil proteins, and these suggested an affiliation to the Denisovans, as did ancient DNA from sediments in Baishiya Cave, its reported place of origin.

ABOVE The robust Xiahe mandible has been linked with Denisovans through its large molars and its fossil proteins.

ABOVE Baishiya Karst Cave is located in Xiahe County, on the northeastern edge of the Tibetan Plateau. The Xiahe mandible was reportedly found in this sacred cave by a Buddhist monk in 1980.

In 2021, a nearly complete cranium at least 146,000 years old from Harbin in northeast China was published, and analyses of its shape and anatomy suggested it could be grouped with the fossils from Dali, Hualongdong, Jinniushan and Xiahe as a distinctive set of early humans. The cranium is huge in size, with an endocranial volume of ~1400 ml. It displays archaic features such as a long and low braincase with a massive browridge, but it resembles *H. sapiens* in features such as the delicate cheekbones, retracted face and large mastoid processes. The upper jaw is very wide, but has only one very large second molar preserved. Some Chinese scientists created the new species name '*Homo longi*' (Dragon Man) for the fossil, but as it matches the Dali cranium, which has sometimes been called *Homo daliensis*, this seems a more appropriate name for the fossil and the Chinese material that resembles it.

ABOVE A facial view of the large and beautifully preserved Harbin cranium from China.

LEFT A whole body reconstruction of the Harbin individual from China. Only the cranium was discovered, but its huge size and robusticity suggest this was a large-bodied adult male.

Thus we have a group of fossils in East Asia that are anatomically distinct from *Homo neanderthalensis* and *Homo sapiens* (*Homo daliensis?*), and we also have a set of genomes in East Asia that are genetically distinct from those of *Homo neanderthalensis* and *Homo sapiens* (the Denisovans) – perhaps they are one and the same? They may well be, but this question cannot be fully resolved until we can combine more complete fossils from the region with recognisable Denisovan genomes.

Homo floresiensis

At the beginning of this millennium, most palaeoanthropologists thought that only modern humans had managed to travel beyond Southeast Asia towards New Guinea and Australia in the past 60,000 years, because sea-going watercraft were essential for such a journey, and it seemed very unlikely that archaic humans had such technology. Excavations that were destined to challenge this idea began in 2001, led by the archaeologist Mike Morwood. In the huge Liang Bua Cave ('cool cave') on the remote Indonesian island of Flores, about 500 km (310 miles) to the east of Java, Morwood's team reopened old trenches several metres deep and soon encountered promising finds such as stone tools more than 10,000 years old, and fossils of a pygmy form of the extinct elephant-like *Stegodon*. In 2003, at a depth of around 6 m (19½ ft), the team encountered a small skeleton, which at first they thought must represent a modern human child. However, the tiny-headed skull showed definite brow ridges above its large eye sockets, and the wisdom teeth had fully erupted, indicating this was an adult individual, not a child.

A year later Morwood and colleagues published controversial analyses proposing that the small skull and skeleton represented a new human species *Homo floresiensis*, but soon nicknamed the hobbit, owing to its diminutive stature of about 1.05 m (3½ ft). They dated the skeleton at less than 20,000 years old, suggesting that a primitive hominin had survived on Flores well into the era of modern humans. Furthermore, they argued that it represented a unique example of insular dwarfism in humans (a process whereby large mammals isolated on islands evolve smaller bodies in response

BELOW Liang Bua, Flores, is the huge cave that has yielded over 100 fossils of the controversial species *Homo floresiensis*.

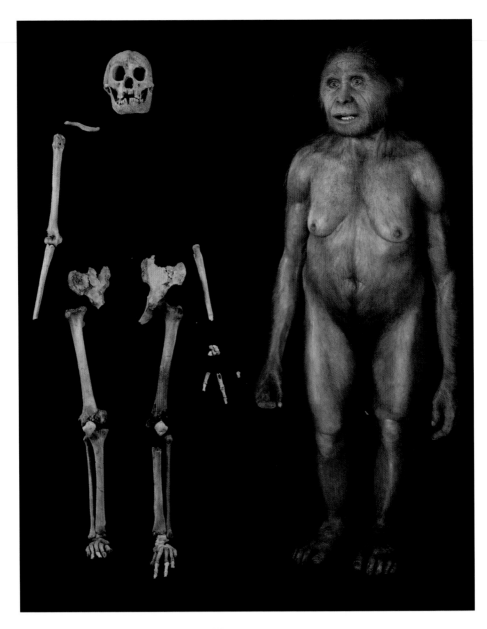

ABOVE The partial skeleton Liang Bua 1, an adult female of *Homo floresiensis* and a reconstruction showing the unusual body shape and large flat feet of *Homo floresiensis*.

to limited resources and the lack of predators). The suggestion was that a population of *H. erectus* could have travelled to Flores from Java, perhaps by boat, and then shrank in size over hundreds of thousands of years to become *H. floresiensis*. In line with a 'human' classification, charcoal and stone tools apparently associated with *Stegodon* bones in Liang Bua suggested that *H. floresiensis* could have used fire, and hunted and butchered these animals in the cave.

Further studies of the original skeleton and fragmentary fossils of up to eight more individuals suggested that features such as broad, flared hipbones, a short collarbone, and a forwardly positioned shoulder joint all resembled australopiths. The tiny brain of the *H. floresiensis* skull (volume about 420 cm^3), its body shape, and its foot, hand and wrist bones look more archaic than those of any human dating to within the past million years, while primitive traits of the wrist bones and jaw are replicated in at least one more individual from the site. A second lower jaw is also small, lacks a chin, and shows internal ridges like those found in pre-*erectus* fossils. Yet the small teeth and aspects of the skull shape do suggest that this could be a diminutive relative of *H. erectus*. The small brain size of *H. floresiensis* has provoked especially fierce controversy with some suggesting that it could derive from a *H. erectus* brain, diminished but human in structural organization, while others have argued that it is abnormal, caused by a condition called microcephaly operating on a modern, not archaic, human.

Mike Morwood died in 2013, but many of the team have continued working at Liang Bua and other sites in Flores. In 2016 they revised their previous conclusions about the dating of the Liang Bua sequence, suggesting that the *H. floresiensis* material and associated stone tools and animal remains actually derived from deposits that are at least 60,000 years old, while related stone tools may have extended for another 10,000 years or so. This new dating undermines continuing claims that the hobbit fossils belong to diseased modern humans, since the material now dates beyond any modern human specimens known from the region. And a last appearance of *H. floresiensis* at about 50,000 rather than 15,000 years ago, brings that event much more in line with the inferred last appearances of the Neanderthals and Denisovans. Moreover, teeth attributed to *H. sapiens* have been recovered from immediately succeeding levels at Liang Bua, dating from about 46,000 years ago, suggesting that modern humans could have produced the evidence of fire in the cave, and might have replaced *H. floresiensis* on the island by that time.

In 2015 the team also published information about fragmentary but much more ancient human fossils discovered at the site of Mata Menge in central Flores, dating to about 700,000 years ago. This site had already produced stone tools attributed to an ancestor of *H. floresiensis*, and even older stone tools were found at the nearby site of Wolo Sege, suggesting an occupation about 1 million years ago. The new fossils consisted of a fragmentary lower jaw and several teeth representing at least three individuals. All the teeth are about the same size or even smaller than those known from Liang Bua, seemingly confirming the existence of the diminutive *H. floresiensis* lineage at least 700,000 years ago, and further countering claims of pathology in modern humans as an explanation for the morphology of *H. floresiensis*.

These continuing finds on Flores hint that there could be many more surprises to come from the rest of Wallacea (the isolated chains of islands lying between Southeast Asia and Papua-New Guinea/Australia). If the ancestors of *H. floresiensis* reached Flores, perhaps they also spread to other islands, and the experiment in human evolution revealed at Liang Bua might have been duplicated elsewhere – for example in the Philippines, where another apparently dwarfed human species was recently named, *Homo luzonensis* (see below). As Morwood pointed out, the powerful currents around Indonesia might have favoured transport from Sulawesi (north of Flores) rather than from Java, where the nearest *H. erectus* fossils have been found. The possibility of accidental rafting on mats of vegetation in such a volcanically active part of the world must also be considered: in the 2004 Indian Ocean tsunami, some people survived on floating debris for a week or more, and ended up about 150 km (93 miles) out to sea. If that could happen in the last few years, what might seem an improbable stranding on Flores looks more feasible when we consider time scales of hundreds of thousands of years.

Homo luzonensis from Southeast Asia

The possibility that there were more surprises to come from the islands of Southeast Asia after the discovery of *H. floresiensis* has been confirmed on the island of Luzon in the Philippines, nearly 3,000 km away. 13 fossil human remains – teeth, hand and foot bones, and part of a femur from at least 3 adult and immature individuals have been recovered from excavations in Callao Cave since 2007. They have been dated to at least 50,000–67,000 years old, are small in size, particularly the teeth, and they show a distinctive combination of primitive and derived traits sufficient for the creation of a new human species: *Homo luzonensis*. Some of the hand and foot bones show features also present in much more ancient australopith species from Africa, where they have been interpreted as adaptations for life in the trees. As with *H. floresiensis* it is difficult to disentangle genuinely primitive traits from those that could have evolved on Luzon, and there are the same discussions about whether such traits might hark back to a dispersal from Africa that happened prior to the appearance of *H. erectus*. Alternatively, as also suggested for *H. floresiensis*, is the Luzon population a descendant of *H. erectus*, which underwent isolation and island dwarfing over a considerable period of time? Archaeological evidence suggests that some form of human on the island was butchering a rhinoceros with stone tools about 700,000 years ago, but how that relates to the early history of *H. luzonensis* is unknown without associated fossil remains. As for the fate of *luzonensis*, it is too early to say whether the spread of *Homo sapiens* into the region at least 50,000 years ago might have been a factor in its extinction, as has also been suggested for the disappearance of *H. floresiensis*.

LEFT The huge Callao Cave on the island of Luzon (Philippines) has produced the remains of the newly-named *Homo luzonensis*.

CHAPTER 6

Homo sapiens

Today, we are the only human species on Earth, and we range more widely geographically than any previous human. Although people around the world come in many different sizes, shapes, and skin colours, with many cultural differences, we undoubtedly all belong to the one species, *Homo sapiens*. Anatomically, living humans share specific features in the skeleton that can also be recognized in fossil remains, such as a high and rounded braincase, a small face tucked under the forehead, a chin on the lower jaw (even in infants), small and separated brow ridges, and relatively narrow shoulders, trunk and pelvis. Data from the three-dimensional X-ray technique known as Computerized Tomography (CT) show that our middle and inner ear anatomies and our microscopic dental structure are similarly distinct from other human species. Discernible by the minute daily growth lines in the enamel of our teeth, which can be used to 'age' an individual, modern humans are also characterized by long and slow growth to maturity, allowing time for the brain to develop, and for cultural knowledge to be absorbed. This extended growth period seems to be a distinctive modern human feature, although nearly matched by the Neanderthals, with their similarly large brains.

OPPOSITE An assemblage of Middle Stone Age flint tools from Jebel Irhoud, Morocco, dated to about 315,000 years old.

Early *Homo sapiens*

While the Neanderthals were evolving in Eurasia, *Homo sapiens* were evolving in Africa. Early examples of this evolution may include the currently undated fossil cranium discovered by tourists visiting Eliye Springs (West Turkana, Kenya), and the better-dated front of a skull and partial face from Florisbad (South Africa), placed at about 260,000 years ago from direct Electron Spin Resonance (ESR) dating (see p. 33) of its molar tooth. The Jebel Irhoud quarry site in southwestern Morocco has provided more complete evidence of an emerging *H. sapiens* anatomy in the form of five individuals represented by three partial or nearly complete skulls, an adult jawbone with most of its teeth, the jawbone of a child aged about seven, and various bones from the rest of the skeleton. These were formerly believed to be about 160,000 years old, but new dating work places them at about double that age. The braincases have strong brow ridges and large brain volumes of 1,300–1,400 cm^3, and their shape resembles both Neanderthals and *H. sapiens*. However, the preserved faces are broad but transversely flat, and most resemble a large version of the *H. sapiens* face. Yet another specimen that combines a rather primitive braincase with a more modern-looking face is the Laetoli 18 cranium, found at Ngaloba in Tanzania. It is often dated in the region of 140,000 years old, which, if correct, would support the view that archaic skull morphologies persisted alongside more modern-looking ones.

LEFT A reconstruction of a European male, decorating his body, based on skeletons from the Gravettian cultural phase, about 30,000 years ago.

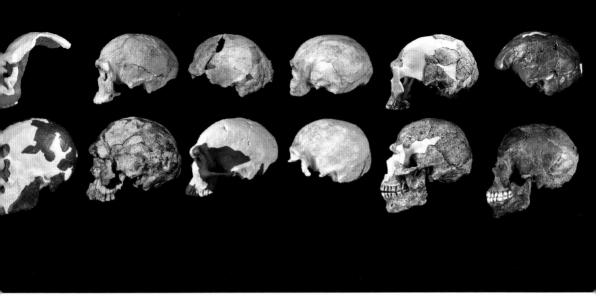

ABOVE An array of early *Homo sapiens* skulls from Africa and Israel, dating between about 300,000 (top left) and 100,000 (bottom right) years ago. The variation shown in size and shape is striking and suggests that the evolution of *Homo sapiens* was a complex and pan-African process, rather than having a single centre of origin.

Modern *Homo sapiens*

The African fossils discussed so far show only some of the elements of the modern skeletal pattern. By 160–230,000 years ago, there were humans much closer to the modern anatomical pattern in Ethiopia, as represented by the partial skeleton Omo Kibish 1, and crania Herto 1 and 2. The skull of Omo Kibish 1 had to be extensively reconstructed, but does display features such as a high and rounded braincase and a chin on the lower jaw. However, to complicate the picture, Omo Kibish 2, a faceless cranium found several kilometres away, shows much less of the modern pattern. While it does show only slight brow-ridge development, and a similarly large brain volume (about 1,400 cm^3), the cranium is angular rather than rounded, with a flexed occipital bone at the back, like *H. heidelbergensis* or *H. erectus* specimens. The Omo 1 skeleton lies under a volcanic deposit that dates from about 233,000 years ago, while Omo 2 was a surface find from a different location that is not so well dated. If they are of similar age, does this show great variability within the population, or does it indicate that quite distinct populations were living in close proximity at that time? Another partial cranium from Guomde (West Turkana), which is of

LEFT A replica of the Omo Kibish 1 partial skeleton from Ethiopia, dated over 230,000 years old. Part of the hip bone found more recently suggests this might have been a female individual, despite its large size.

similar or somewhat greater age, combines some of the features of both Omo Kibish 1 and 2, and a modern-looking thigh bone was discovered nearby.

So 200,000 years ago humans in Africa must have been a very diverse group, and some began to spread beyond the continent around that time. The back of a *sapiens*-like cranium from Apidima Cave in Greece, has been dated to to over 200,000 years old, and a small-toothed upper jaw was excavated from Misliya Cave in Israel and dated to around 170,000 years. Further early modern humans were present in western Asia around 120,000 years ago, apparently taking advantage of the relatively warm and wet conditions in the region then. In the 1930s, the small cave of Skhul in what is now Israel produced the fossils of 10 adult and immature early *H. sapiens* individuals, possibly representing a cemetery used over a period of time. One of the males was buried clasping the massive lower jaw of a wild boar. Soon after those excavations, work at the Israeli cave of Qafzeh began, eventually producing a similarly large sample of early *H. sapiens* remains, including the burial of a child covered with the large antlers of a deer. Both sites also contain evidence of imported natural red pigments and sea shells, apparently pierced to form jewellery. Nevertheless, the stone tools found with these materials are clearly Middle Palaeolithic and not notably different from those found with Neanderthal remains in western Asia. The Skhul and Qafzeh skeletons may or may not represent the same kind of people – we cannot date them accurately enough to tell whether they are in fact spread out over many thousands of years. They show a rather primitive, but fundamentally modern anatomy; large and predominantly rounded braincases that nevertheless often have strong brow ridges, large but modern-looking faces, jawbones with variable development of a chin, variably sized teeth and a strongly built but tall and linear physique, contrasting markedly with the Neanderthals. Adult body-weight and height is estimated at about 63–77 kg (139–170 lb) and 1.70–1.90 m (5½–6¼ ft).

It is generally acknowledged from genetic data that the main dispersal of *H. sapiens* from Africa took place only about 60,000 years ago, so how do the primitive modern humans from Skhul and Qafzeh, dating from around 100,000 years ago, fit into this pattern? They have usually been considered as, in a sense, part of a failed dispersal from Africa that went no further, and died out. However, recent data suggest that this is not necessarily the case. Firstly, there is archaeological and fragmentary fossil evidence of modern humans in Arabia at least 90,000 years ago, and suggestions that early modern populations in Arabia dispersed into India by at least 75,000 years ago. Additionally, many isolated modern human teeth have been recovered from the Fuyan Cave in southern China in 2011–2013, and these have been dated, somewhat controversially, to at least 80,000 years in age. Possibly supporting finds come from Sumatra, where two modern human teeth have recently been dated to about 70,000 years, and there are claims that a single modern human tooth from Punung in Java dates to about 120,000 years. In addition, as mentioned later, there is possible archaeological evidence that modern humans even reached northern Australia by 65,000 years ago. All of this suggests that early modern humans could have extended across southern Asia at least 80,000 years ago, but given that the strongest genetic signal in living humans outside of Africa indicates only a dispersal about 60,000 years ago, it may well be that early migrations did extend further east than western Asia, but these populations were supplanted by the later wave.

Interestingly, there is evidence from Neanderthal genomes of an early interbreeding event with modern humans placed more than 200,000 years ago, although the exact location and date of the event cannot be determined. It may

BELOW A reconstruction of the Mungo 3 burial in southern Australia, about 42,000 years ago.

well be that the Apidima 1 fossil represents an early *H. sapiens* population that contributed to this interbreeding with Neanderthals.

There is an unfortunate lack of modern human fossils in the time window of 40,000–60,000 years ago that could help map the dispersal of *H. sapiens* during that period. Further below (see p. 150) we discuss fossils from Siberia, Czechia, Bulgaria and Romania that date from about 45,000 to 40,000 years ago, but otherwise clues are scattered and sparse. They include an isolated child's tooth from Grotte Mandrin in southern France about 54,000 years old, a small-sized modern human partial cranium from Manot Cave in Israel dating from about 55,000 years ago, and a similarly aged modern-looking skull and separate, much more strongly built lower jaw from Tam Pa Ling in northern Laos. Tracing the dispersal of modern humans further south, there is a partial cranium from Niah Cave, Sarawak, Malaysia, dating from about 40,000 years ago, indicating that humans were already adapting to life in the local rainforests by that time, and paintings of wild pigs in caves in Borneo and Sulawesi from around the same time. Also of similar age are two very large but modern-looking crania with some other bones from Wajak Cave, Java. Interestingly, these were recovered by Eugene Dubois while he was searching for what would become the first known fossils of *H. erectus* on the island. Finally, in southeastern Australia, alongside large and now arid lake basins, there are the continent's oldest fossils, the cremated remains of Mungo 1 and the partial skull, jaw and skeleton of Mungo 3. Mapping the furthest known extent that modern humans had reached by 42,000 years ago, these constitute the oldest known cremation and one of the oldest known burials containing red ochre pigment. Controversial evidence of an even earlier arrival of modern humans in Australia comes from the archaeological site of Madjedbebe 2 in Arnhem Land, where stone tools and extensive accumulations of pigment have been dated to about 65,000 years in age.

By about 50,000 years ago, modern humans were beginning to penetrate the European territories of the last Neanderthals. Why they took so long to do so is an unsolved mystery – were they there much earlier but as yet undetected, did they have to develop extra cultural adaptations to cope with European environments, or were the Neanderthals able to out-compete them? Whatever the answer, *H. sapiens* eventually took root in Europe and grew in numbers as the Neanderthals vanished. These early modern Europeans are sometimes called Cro-Magnons after 30,000-year-old remains discovered in a French rock shelter in 1868. In fact, a similarly aged partial skeleton stained with red

HUMAN EVOLUTION DURING THE LAST 1 MILLION YEARS

As is evident from this book, our understanding of human evolution is constantly evolving. In the diagram opposite we have tried to summarize the inferred age ranges of hominin lineages during the last million years, but this is, of course, only one interpretation of the species and their evolutionary relationships. Colours reflect the species designations used in this book and dotted lines show putative ancestor-descendant relationships and possible continuations through time. The diagram illustrates the growing evidence of the contemporaneity of distinct human lineages (we would argue species) at every level of the diagram until the last 40,000 years or so.

The evolutionary history of *H. naledi* is particularly enigmatic, confounded by its unexpected combination of traits that makes it difficult to establish whether the species diverged early on in the evolution of the genus *Homo*, or split off from another lineage more recently.

Homo sapiens is shown with an 'archaic' or basal branch, representing fossils like Jebel Irhoud and Omo Kibish 2, showing only a minority of modern or derived traits such as globular braincase, small brow ridge etc. However, some of these fossils may lie on the ancestry of later *H. sapiens*, and Africa is probably the place where a 'braided stream' analogy of recent human evolution is most appropriate.

Moving across the diagram, *H. heidelbergensis/ rhodesiensis* is shown as a separate branch terminating about 300,000 years ago, but its last appearance is unknown.

Genetic data suggest the Neanderthal-Denisovan divergence occurred about 400,000 years ago, but as depicted here, it probably preceded early Neanderthals as represented by the Sima de los Huesos fossils, now dated to about 430,000 years ago, and placed on the Neanderthal lineage through anatomy and nuclear DNA.

The exact classification of fossils like Harbin, designated here as *H. daliensis* (alternatively known as '*H. longi*'), is still unclear, but they may be equivalent to the Denisovans, who are still mainly known from their DNA. A more complete picture should emerge when Denisovans genomes are known from more complete fossils.

H. floresiensis (shown here including earlier fossils from Flores) may be a descendant of *H. erectus* that dwarfed on the island of Flores, or alternatively it may have evolved from an even deeper, pre-*erectus* divergence. The origins of *H. luzonensis* are similarly obscure at the moment.

While *H. floresiensis* survived into the last 60,000 years, our representation of the extinction dates for the Asian and Java *H. erectus* branches is speculative in the absence of better dating of these events. We have depicted *H. antecessor* as a short-lived branch, but we know next to nothing about the distribution of this species in time and space.

Evidence indicates that there was gene flow (red lines) at various times throughout the last 1 million years, although the rate and frequency of this before about 100,000 years remain to be established.

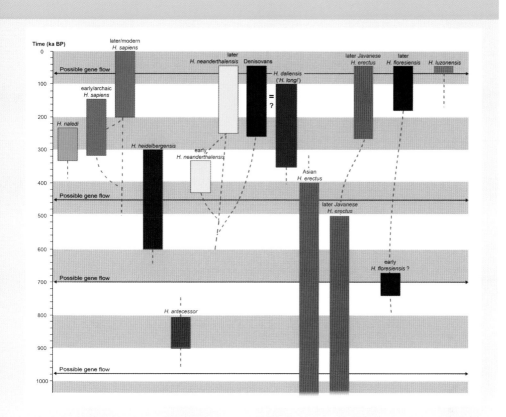

ABOVE This representation of human evolution during the last 1 million years shows the main lineages and species, together with some possible connections between them.

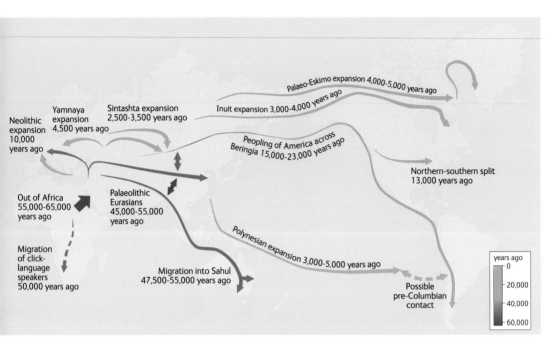

ABOVE Genetic data complement fossil and archaeological evidence to map the spread of *Homo sapiens* across the Earth. They do not always agree, however. Recent archaeological research suggests human arrival in Australia by 65,000 years ago, while footprints in Mexico and artefacts from several sites in the Americas suggest a human arrival there more than 23,000 years ago.

pigment and buried alongside mammoth ivory jewellery had been excavated 45 years earlier at Paviland Cave in south Wales, but its significance was not understood at the time.

The Europeans skeletons associated with Upper Palaeolithic industries between about 11,000–40,000 years ago show a great deal of variety, with ones from before the peak of the last ice advance (about 20,000 years ago) having taller and more linear body shapes, reminiscent of some modern East Africans, whereas those from after the last glacial maximum tend to be shorter and stockier, with an estimated body-weight and stature in the range of 52–84 kg (114½–185 lb) and 1.60–1.80 m (5¼–6 ft). On average Upper Palaeolithic people were longer-headed, larger-brained (values ranging from about 1,200–1,800 cm³) and larger-toothed than present-day Europeans, and their DNA shows that they were relatively dark in complexion (skin, eye and hair colour). It seems that blue eyes, fair and red hair, and a light complexion only started to become common in Europe during the last 15,000 years.

Outside of Europe, modern human skeletons between 12,000–40,000 years old are relatively rare, but where they exist in Africa, the Far East, Australasia and the Americas, they indicate that skeletal variation in *H. sapiens* was previously much greater than we find today. For example, around 10,000–15,000 years ago, several samples show evidence of a greater robusticity in the skull. This is true for fossils found in Nigeria (the Iho Eleru partial skull and lower jaw), Kenya (West Turkana), China (Maludong Red Deer Cave and Longlin Cave) and Australia (Kow Swamp). This has led to suggestions that some of this greater variation may be due to a larger amount of persisting 'archaic' genes from previous interbreeding events (see p. 147).

Until recently it was thought that the Americas were reached by modern humans only within the last 20,000 years or so. It was assumed that a significant constraint on this migration from Asia was the need to adapt to sub-Arctic conditions before reaching far enough north to access either the Bering Strait land bridge across to Alaska, or the chains of islands to the immediate south. There is now evidence that an early wave of modern human colonizers managed to reach as far as Alaska by about 30,000 years ago, while artefacts from several sites and footprints in Mexico suggest a movement southwards before 23,000 years ago.

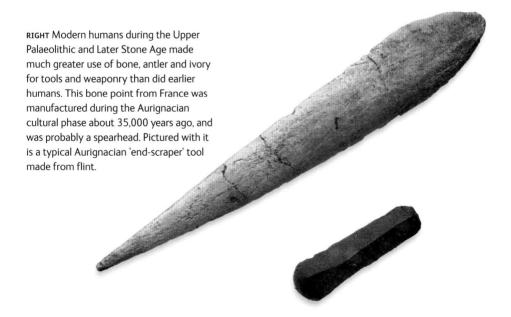

RIGHT Modern humans during the Upper Palaeolithic and Later Stone Age made much greater use of bone, antler and ivory for tools and weaponry than did earlier humans. This bone point from France was manufactured during the Aurignacian cultural phase about 35,000 years ago, and was probably a spearhead. Pictured with it is a typical Aurignacian 'end-scraper' tool made from flint.

Late *Homo sapiens*

What enabled the successful spread of modern humans across the inhabited world, and what happened to the Neanderthals and the other surviving archaic humans outside of Africa? Fifteen years ago many scientists thought that the spread of modern humans from Africa, about 60,000 years ago, led to a total replacement of the other human groups on Earth, give or take a minor amount of mixing, which would not show up in people today. Palaeoanthropologists generally regard forms like the Neanderthals as distinct species from us, with separate evolutionary pathways over hundreds of thousands of years, but this does not necessarily mean that the gulf was wide enough to prevent some level of interbreeding, since closely related mammal and bird species today may still be able to hybridize, even after a million or more years of evolutionary separation.

And indeed, in the last decade or so we have learnt that humans originating outside of Africa carry around 2% Neanderthal-derived DNA in their genomes, from interbreeding events that are believed to have happened 40,000–60,000 years ago. In addition, some populations in the islands of Southeast Asia and Oceania (especially the Philippines, Papua New Guinea and Australia) carry an even higher level of DNA in their genomes from ancient interbreeding with a Denisovan-like population. Since the ancestors of Southeast Asians and Oceanians are thought to have migrated through southern and then Southeast Asia rather than venturing anywhere near Siberia, this indicates that there was probably a Denisovan-like population living in Southeast Asia as well as Siberia, and the interbreeding occurred there.

There is also some genetic evidence that present-day sub-Saharan African genomes carry signs of interbreeding (also known as introgression) from an archaic human species such as *H. heidelbergensis* or *H. naledi*. While most of the introgressed DNA in our genomes seems to be non-functional, some elements could have provided health benefits in the past such as faster blood-clotting and resistance to infectious diseases, since archaic humans outside of Africa would have developed immunities to local pathogens, and passed those on in interbreeding events. However, there may also be negative effects today in terms of an increased risk of strokes and in some auto-immune conditions. Some early modern fossils provide clues from their DNA on when and where interbreeding took place. For example, study of the lengths of strands of Neanderthal DNA in a fossil leg bone from Siberia (Ust'-Ishim) and a partial skeleton from Czechia

(Zlatý kůň), both dated to about 45,000 years ago, indicates that their Neanderthal DNA was acquired at least 5,000 years earlier, while DNA in teeth from Bulgaria (Bacho Kiro) about 44,000 years old and a lower jaw from Romania (Oase) dated to about 40,000 years ago suggest these individuals had a Neanderthal ancestor within the previous 200 years or so. This may also be true for teeth from La Cotte de St Brelade in Jersey, Channel Islands, which show mixed Neanderthal and *H. sapiens* traits, although in this case there is no ancient DNA so far to confirm that they are evidence of interbreeding.

Finally, we need to consider why *H. sapiens* is the only surviving species out of the many we have discussed above. One possibility is that modern humans simply out-competed the other species where they overlapped in time and space, through more effective hunting and gathering, cultural adaptations, and the occupation of the most favourable environments. Moreover, the genetic data show that at least some of the other human populations were low in numbers and diversity at the time of the modern human dispersal from Africa, and thus these populations were perhaps especially vulnerable to extinction. Yet although the replacement of peoples like the Neanderthals and Denisovans may have happened relatively quickly, it was not an overnight event, and genetic data show that it was certainly not a total replacement in all the regions outside of Africa. To better understand the reasons for our eventual success we are not only going to require much more genetic data from the fossils and from cave sediments, but also a far richer fossil and behavioural record from regions like Asia and Southeast Asia, as well as Africa. This is especially apparent when we consider a completely unexpected find like *H. floresiensis* from a remote island in Indonesia.

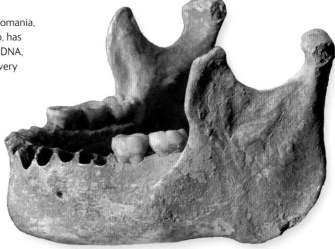

RIGHT The Oase mandible from Romania, dated to about 40,000 years ago, has about a 9% level of Neanderthal DNA, suggesting this individual had a very recent Neanderthal ancestor.

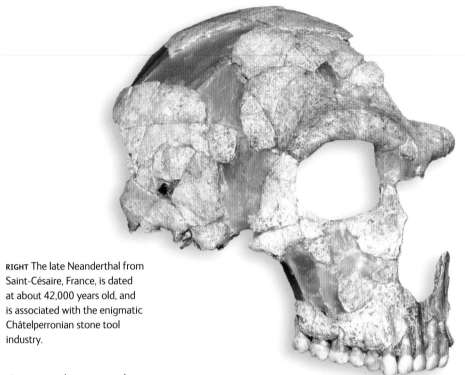

RIGHT The late Neanderthal from Saint-Césaire, France, is dated at about 42,000 years old, and is associated with the enigmatic Châtelperronian stone tool industry.

A complex puzzle

'Light will be thrown on the origin of man and his history'. These were Charles Darwin's prophetic words in 1859, at the end of his book *On the Origin of Species*. But how much of our evolutionary history is now illuminated, and are the overall patterns becoming clear? The confident view of several leading palaeoanthropologists in the 1970s and 1980s was that the basic story had been told, as many important discoveries of australopith and early *Homo* species were being made in Africa. However, as readers of this book should now be aware, further discoveries, better dating and the advent of ancient DNA studies have shown that the confidence of those palaeoanthropologists was, in fact, over-confidence. The simple linear sequence of *Au. afarensis* giving rise to *H. habilis*, which in turn evolved into *H. erectus*, which eventually gave rise to *H. sapiens*, seemed quite defensible when most of the fossils could apparently be grouped into these broad categories. But now, within each of those 'stages' of human evolution, we can see a diversity of species, with the pattern forming a radiating bush of foliage, rather than a ladder of progress leading irrevocably to *H. sapiens* at the summit.

As we have seen, in the earliest part of the record there are several candidate species for the earliest members of the hominin line, after our ancestors diverged from those of the chimpanzees. While each of them – *Sahelanthropus*, *Orrorin* and *Ardipithecus* – possess hominin characteristics such as small canine teeth and signs of being bipedal, is the evidence complete enough or convincing enough to confirm their place in the human lineage? Once we arrive at about 4 million years ago, the earliest species of *Australopithecus* are much more convincing as habitual bipeds, reinforced by the evidence of the Laetoli trackways. Yet in many other ways, these creatures seem predominantly ape-like in their growth, diets and behaviours, while their upper body anatomy reveals that they had not fully abandoned life in the trees. Which of the many species we have described might have given rise to the first true humans (genus *Homo*) in the time period 2–3 million years ago, is quite uncertain, and isolated discoveries from regions like Chad remind us what a biased record of human evolution we still have in Africa. Dominated by the sedimentary basins of East Africa and the limestone caves of South Africa, our finds come from only about 10% of the area of the continent.

When we turn to early *Homo* itself, the pattern remains equally complex, with a probable co-existence 1.8–1.9 million years ago of fossils assigned to *H. habilis*, *H. rudolfensis* and *H. erectus*, not to mention the near-contemporaneous *Au. sediba* in South Africa, which also showed some human-like traits in its skeleton. By 1.5 million years ago, it seemed that the other primitive *Homo* forms were vanishing as *H. erectus* thrived, and the robust australopiths soon followed them to extinction. Yet we must now recognize the remarkable survival of *H. naledi* into the last 400,000 years, with its unique mixture of traits reminiscent of early *Homo* and even australopith skeletons. If those primitive features are indicators of an ancient and persisting heritage from at least 2 million years ago, then *H. erectus* cannot have been alone in Africa a million years ago.

We then move on to human evolution outside of the ancestral homeland, with the type fossils of all the other generally recognised species apart from *Homo sapiens* – *H. erectus*, *H. antecessor*, *H. heidelbergensis*, *H. neanderthalensis*, *H. daliensis*, *H. luzonensis* and *H. floresiensis* – being found outside of Africa. The prevailing view that *H. heidelbergensis* evolved from *H. erectus* (region of origin unknown) may still be true, but that species' position as the 500,000-year-old common ancestor of Neanderthals and modern humans is currently under reconsideration from both fossil and genetic data, with the possibility of a more ancient common ancestor, possessing a face more like that of *H. antecessor* than

H. heidelbergensis. Here, too, the old idea of a neat succession of species has been eroded by evidence of an overlap of *H. heidelbergensis* (or *rhodesiensis*?) with both early *H. neanderthalensis* and *H. sapiens* (and now with the addition of *H. naledi*). This is followed by a subsequent co-existence of Neanderthals, *H. sapiens*, Denisovans and *H. floresiensis*, with the lingering uncertainty of a possible late survival of *H. erectus* in regions like Java as well. Finds like *H. floresiensis* and the Denisovans serve to remind us that it is not only Africa that is poorly sampled for human fossils. Ancient human diversity in Asia and Southeast Asia is arguably even less well-represented by the current fossil record.

Several workers have suggested that the evidence of interbreeding now apparent between (at least) Neanderthals, modern humans and the Denisovans, means that we should regard all of these as members of our own species *H. sapiens*, as they were all interfertile. However, this ignores the fact that the 'biological species concept' does not even work well when applied to many generally recognized species of birds and mammals today, which often share at least small amounts of DNA with their closest relatives. So its application to fossil lineages is bound to be even more problematic. If clearly diagnosable evolutionary lineages can be recognized through hundreds of millennia, as we would argue for *H. neanderthalensis* and *H. sapiens*, then in our view a small amount of hybridization between them should not override their classification as distinct species. But there is no doubt that the level of hybridization between different hominins that we have learnt about in the last few years is only a minimum, and realistically, it was probably a feature of human evolution throughout much of the last 7 million years.

Finally, what can we say about our own place in the story, as the last survivors of a complex evolutionary radiation? Looking at the early parts of the history of our species, there was nothing inevitable or preordained about our success, and yet the fact that Neanderthals, Denisovans, *H. floresiensis*, *H. luzonensis* and perhaps also *H. erectus* all went physically extinct in the last 70,000 years as modern humans grew in numbers and range, seems to us more than a coincidence. *Homo sapiens* may not have aimed to cause those human extinctions, just as we do not mean to cause the extinction of the Pyrenean ibex, river dolphins, and (soon?) our great ape relatives, but our increasingly global spread seemingly came at the price of losing the physical (if not genetic) traces of our closest relatives. We hope our book has succeeded in illuminating those lost vestiges of humanity who once shared the planet with us, and at an earlier time, with our African ancestors.

Further reading

Almécija, S., Hammond, A.S., Thompson, N.E., Pugh, K.D., Moyà-Solà, S., Alba, D.M., Fossil apes and human evolution. *Science*, 2021, 7: 372(6542).

Bergström, A., Stringer, C., Hajdinjak, M., Scerri, E.M.L., Skoglund, P., Origins of modern human ancestry. *Nature*, 2021, 590(7845): 229–237.

DeSilva, J., *First Steps: How Walking Upright Made Us Human*. HarperCollins Publishers, 2021.

Dinnis, R. and Stringer, C., *Britain: One Million Years of the Human Story*. The Natural History Museum, 2014.

Graeber, D., Wengrow, D., *The Dawn of Everything: A New History of Humanity*. Penguin/Allen Lane, 2021.

Henke, W. and Tattersall, I., *Handbook of Palaeoanthropology*. Springer, 2015.

Lieberman, D., *The Story of the Human Body: Evolution, Health and Disease*. Penguin, 2013.

Higham, T., *The World Before Us: How Science is Revealing a New Story of Our Human Origins*. Penguin/Viking, 2021.

Leakey, M., Leakey, S., *The Sediments of Time: My Lifelong Search for the Past*. Houghton Mifflin Harcourt, 2020.

Newson, L., Richerson, P., *A Story of Us: A New Look at Human Evolution*. Oxford University Press, 2021.

Pääbo, S., *Neanderthal Man: In Search of Lost Genomes*. Basic Books, 2015.

Papagianni, D. and Morse, M.A., *The Neanderthals Rediscovered: How Modern Science Is Rewriting Their Story*, 2nd ed. Thames and Hudson, 2015.

Pattison, K., *Fossil Men: The Quest for the Oldest Skeleton and the Origins of Humankind*. HarperCollins Publishers, 2020.

Reader, J., *Missing Links: In Search of Human Origins*. Oxford University Press, 2011.

Reich, D., *Who We Are and How We Got Here: Ancient DNA and the new science of the human past*. Oxford University Press, 2018.

Ungar, P.S., *Evolution's Bite: A Story of Teeth, Diet, and Human Origins*. Princeton University Press, 2017.

Wragg Sykes, R., *Kindred: Neanderthal Life, Love, Death and Art*. Bloomsbury Publishing, 2021.

Wood, B., *Wiley-Blackwell Encyclopedia of Human Evolution*. Wiley-Blackwell, 2011.

Wragg Sykes, R., *Kindred: Neanderthal Life, Love, Death and Art*. Bloomsbury Publishing, 2020.

Index

Pagination in *italic* refers to illustration captions. **Bold** pagination refers to in-depth features.

Acheulian handaxes 89
adolescents, fossils of 45, *46*, 92, 115
Afar region, Ethiopia 52, 65, 68, *83*
age determination 56, 139
AL 666-1 jaw *83*
Alaska, USA 149
Alemseged, Zeresenay 55
Algeria
    Tighennif 109, 122
Allia Bay, Kenya 49
Altai Mountains, Siberia 113, 126, 127
Altamira Cave, Spain 93, *93*
Americas, migration to 149
animal carcasses, processing *see* butchery
Anthropocene 33
apes
    differences from hominins 9–14
    see also bonobos; chimpanzees; gorillas *and* orangutans
Apidima Cave, Greece 143
    Apidima 1 fossil 145
Arabia 143
Arago, France 110
Arambourg, Camille 73
Aramis, Ethiopia 27, 31
Ardi 27, 29, *29*
Ardipithecus 21, 153
    Ar. kadabba 18, 31
    Ar. ramidus 18, 27–31, 64
argon-argon dating 33
arms 24, 50, *52*, *53*, 55, 80, 86, 106
    forearm *40*, 67, 125
    upper arm 27, *39*, 67
    wrist 29, 51, 55, 69, 135
    see also hands
Asfaw, Berhane 66
Atapuerca, Spain 104–5, 106, *106*, 115, *115*, 122, 123
Aurignacian cultural phase 91, 92, *149*
Australia 92, 121, 123, 132, 144, *144*, *148*, 150
    Kow Swamp 149
    Madjedbebe 145
Australopithecus 29, 35–41, 48–9, *37*, 38, 65, 73, 81, 153
    Au. afarensis 7, *11*, 19, *35*, 41, 44, 50, 52–61, 64, 66, *83*, 152
    Au. africanus 12, 19, 41–4, 66

Au. anamensis 19, 31, 38, 49–51, *51*
Au. bahrelghazali 19, 61–2, 66
Au. deyiremeda 64, *65*, 65–6
Au. garhi 19, 66–7, *67*, 69
Au. prometheus 41
Au. sediba 19, 44–8, 94, 153

Bacho Kiro Cave, Bulgaria 151
Baishiya Karst Cave, China 129, *130*
Barlow, George 75
Belgium 113
Berger, Lee 46, 101, 105
Berger, Matthew 45
Bering Strait 149
Biache, France 124
Bilzingsleben, Germany 109
biochronology 22–3, 37
bipedalism 7, 9, 23, 26, 29, *29*, 30, 31, 35, 41, 44, 47, *50*, 51, *52*, 54, 59, 64, 84, 153
Black Skull 73, *74*
blood-clotting 150
Bodo, Ethiopia 110
body adornment 90
body decoration 92, 93, *140*
body measurements *see* height *and* weight
bonobos 7, 27, 60
Borneo 145
Bouri, Ethiopia 66, 68, 69, *69*
Boxgrove, England 110, 111, *111*, 112, *112*, 119, 122
brain size *see* cranial capacity
breccia 14, 36, 37, 38, 39, 41, 44, 45
Broken Hill mine, Zambia 109, 110, *110*, 112
Broom, Robert 44, 75
Brunet, Michel 21, 61
Bruniquel Cave, France 91
Bulgaria 145
 Bacho Kiro Cave 151
burials 14, 91, 92, 143, *144*, 145, 148
Burtele, Ethiopia 64
Burtele foot 64, *64*, 65
butchery 17, *17*, 67, 68, *69*, 81, 89, 90, 106, 111, *112*, *126*, 135, 137

Callao Cave, Philippines 136, *137*
canine premolar honing complex 11, 27, 31, 44, 54

cannibalism 17, 106
cave art 84, 90, 92–3, *93*, 145
cave systems, collapsed 14, 35–8, 45, 75, 77
centre of gravity 9, *29*, 54
Ceprano, Italy 110
Chad *24*, 66, 153
 Djurab desert *21*, 22
 Koro Toro 61
 Toros-Menalla 22
Chad, Lake 61
Châtelperronian stone tool industry 93, *152*
chewing 16, *17*, 55, 71, 79, 81; *see also* muscles, chewing
childhood growth period 56, 85, 139
children 91, 100, 106
 Denny 129
 Selam 55–7, *57*
 Taung child 41–3, *43*
 Turkana Boy 96, *98*
chimpanzees 7, 8, 9, *10*, *11*, *12*, 13, *13*, 16, 21, *21*, 27, *29*, 30, 51, 55, 60, 68, 81
China 100
 Baishiya Karst Cave 129
 Dali 129, 131
 Fuyan Cave 144
 Harbin 131, *131*, *132*, 146
 Hualongdong 129, 131
 Jinniushan 129, 131
 Longlin Cave 149
 Maba *128*, 129
 Maludong Red Deer Cave 149
 Shangchen 96
 Xiahe 129, 131
 Xuchang 129
 Xujiayao 129
 Yunxian 110
 Zhoukoudian Cave 97, *99*, 122
Clacton, England 120
Clarke, Ron 38
climates, ancient **121–3**, 126
climbing, adaptations for 13, 27, 29, 35, 40–1, 44, 47, *52*, 55, 57, 64, *64*, 85, 137, 153
clothing 90, 92
coexistence 19, 50, 66, 81, *128*, 153, 154
collarbones 45, 60, 106, 135
common ancestor 9, 13, 73, 85, 109, 112–13, 118, 153
Cooper's D, South Africa 75
Coppens, Yves 52, 73

Cradle of Humankind World Heritage Site, South Africa 35, *36*, 71
cranial capacity 11
 up to 800 cm³ 24, 40, 44, 47, 52, 66, 73, 77, 86, 96, 100, 101, 135
 800 cm³ or above 96, 112, 118, 120, 125, 140, 141, 148
Crayford, England *89*
cremations 145
Croatia
 Krapina 124
Cro-Magnons 92, 93, 145
CT scanning 23, 27, 55, 56, 139
Cueva Mayor cave system, Spain 115
cut-marked bones 17, *17*, 67, 68, *69*, 78
Czechia (Czech Republic) 145
 Dolní Věstonice 92
 Předmostí 92
 Zlatý kůň 151

Dali, China 129, 131
Dart, Raymond 41
Darwin, Charles 7, 94, 152
dating fossils **32–3**, 37, 45, 106
Dear Boy 79
death, causes of 105
 drowning 55, *57*
 falling into caves 14, 39, 45, 46
 injuries 46, 117
 predation 14, 42–3, 51, 77, *77*, 80, 81, 105
Denisova Cave, Siberia 123, *127*, *127*, *128*, 129
Denisovans 8, 18, 123, 127–9, *130*, 132, 135, 154
Denny 129
dentition *see* teeth
diet *30*, 31, 66, 71, 81, 114
 reconstruction 16–17
 and teeth 16–17, *17*, 48, 51, 55, 62, 75, 79
digital reconstruction *43*
Dikika, Ethiopia 55, *56*
Dinaledi Chamber, South Africa 104, 106
disability 92
disease resistance 150
dispersal *see* migration
diversity of species 30, 38, 65, 94, 143, 151, 152, 154
Djurab desert, Chad *21*, 22

Dmanisi, Republic of Georgia 96, 97, 99, 100, *100*, 101
Skull 5 *100*
DNA and genomes *see* genetics
Dolní Věstonice, Czech Republic 92
Drimolen, South Africa 69, *71*, 75, *76*, 101
Dubois, Eugene 94, 145
dwellings 92

earliest hominins 18, *19*, 21–31, 153
ears 118, 139
East Turkana, Kenya 86, *87*, 96, 100
Ehringsdorf, Germany 124
Elandsfontein, South Africa 110
electron spin resonance 33, 140
Eliye Springs, Kenya 140
England
 Boxgrove 110, 111, *111*, 112, *112*, 119, 122
 Clacton 120
 Crayford 89
 Happisburgh *108*, 109
 Swanscombe 118, *119*, 120, *120*, 123
Ethiopia *35*, 52, *64*, 67, 79
 Afar region 52, 65, 68, *83*
 Aramis 27, 31
 Bodo 110
 Bouri 66, 68, 69, *69*
 Burtele 64
 Dikika 55, *56*
 Gona 30, 31, 68, 69
 Hadar 52, *83*, 88, 101
 Ledi-Geraru 88
 Middle Awash 27, 31, 66
 Omo 73, 141
 Woranso-Mille 50, *51*, 60, *60*, 64, 65
evolution, human 94, **146–7**, 152–4
 divergence from apes 7, 8, *8*, 9, 21, *21*, 22, 23
 parallel evolution 73, 85, 94
Excalibur *117*
extinctions 31, 71, 123, 126, 137, 146, 153, 154
eye colour 149

facial projection 9, *12*, 50, 52, 62, 66, 71, 75
feet 30, 31, 47, *88*, *134*, 136, 137

big toe 29, 38, 54, 57, 59, 64, *64*
female hominins
  fossils 27–9, 40, 52, *53*, 56, *57*, *76*, *83*, 88, 126, *134*, *142*
  reconstructions *52*, *57*, *113*, *134*
fire, use of 92, *97*, 135
fishing 92
Flores, Indonesia 123, 132–6
Florisbad, South Africa 140
fluctuations in climate **121–3**, 126
flutes 91, *91*
foods *see* diet
foot bones *see* feet
footprints
  Happisburgh, England *108*, 109
  Laetoli, Tanzania 57–9, *59*
  Mexico *148*, 149
foramen magnum 23, *23*, 44
forensic techniques *43*
fossil assemblages, formation of 81
fossil dating techniques **32–3**, 37, 45, 106
fossil hominin sites (maps) *24*, *36*, *95*, *114*; (timeline) **122–3**; *see also under named sites*
fossil proteins 129, *130*
fossil record, gaps in 15
fossilization process **14**
France 92, 145, *149*
  Arago 110
  Biache 124
  Bruniquel Cave 91
  Grotte Mandrin 145
  Lascaux 93
  Le Moustier 90
  Paris 90
  Saint-Acheul 89
  Saint-Césaire *152*
Fuyan Cave, China 144

Geißenklösterle Cave, Germany *91*
genetics
  DNA and genomes **8**, 93, 113, 117, *117*, 126, 127, *127*, *128*, 129, 132, 144, 146, 148, 150, 151, *151*, 152, 154
  mapping human migration 143–4, *148*, 151
Georgia, Republic of 101
  Dmanisi 96, 97, 99, 100, *100*, 101

Germany 123
  Bilzingsleben 109
  Ehringsdorf 124
  Geißenklösterle Cave *91*
  Heidelberg 109
  Lehringen 90
  Mauer 109, 110, 122
  Neander Valley 113, 124
  Steinheim 118, 124
Gibraltar 113, 114
Gladysvale, South Africa 43
Gona, Ethiopia 30, 31, 68, 69
gorillas 7, 9, 13, 30, 44, 60, 81
Gran Dolina, Spain 106, *106*, 109, 115
Gravettian phase 92, *140*
Greece
  Apidima Cave 143
  Petralona 110
grip, precision 14, 47, *47*, 51, 69
Grotte Mandrin, France 145
Gulf Stream 121
Guomde, Kenya 141

habitats
  forest and woodland 16, 27, *30*, 31, 48, 61, 62, 66, 75, 99, 114, 145
  grassland 25, 61, 99
  lake and riverside 25, 27, 31, 62, 66
  savannah 16, 25, 48, 61, 62, 66, 75, 79
  steppe 113, *121*
Hadar, Ethiopia 52, *83*, 88, 101
Haeckel, Ernst 94, 95
Haile-Selassie, Yohannes 31, 65
hair colour 149
handedness 111, 118
handling, precision 14, 69
hands 13–14, *40*, *47*, 55, 80, 136, 137
  fingers 13, 29, 47, 57, 69, 101, 106, 127
  thumb 47, 69, 101
Happisburgh, England *108*, 109
Harbin, China 131, *131*, *132*, 146
heat, body 125
Heidelberg, Germany 109
height *29*, 60, 95, 99, 101, 112, 118, 124, 133, 143, 148
hips 95, 101, 118, 135, *141*; *see also* pelvis
the hobbit 133, 135

Holocene 33, 121
hominins 7, *18–19*
  differences from apes 9–14
  earliest 18, *19*, 21–31, 153
  species recognition 15–16
  see also under individual species
Homo 35, 38, 66, 69, *71*, 81, 83–137, 152–4
  H. antecessor 18, 106, *106*, 109, 113, 115, 118, 122, 146
  H. daliensis 18, 129–32, 146
  H. erectus 7, *17*, 18, 85, 86, 90, 94–101, 103, 104, *104*, 106, 109, 122, 123, 129, 135, 136, 137, 145, 146, 152
  H. ergaster 101
  H. floresiensis 19, 85, *105*, 123, 132–6, 137, 146, 151
  H. georgicus 100
  H. habilis 18, *78*, 79, 80, 85–94, 101, 103, *104*, 152
  H. heidelbergensis 18, 90, 109–13, 117, 118, 122, 124, 129, 146, 150
  H. longi 131, 146
  H. luzonensis 19, 136–7, *137*, 146
  H. mauritanicus 109
  H. naledi 19, 69, 85, 101–6, *105*, 123, 146, 150
  H. neanderthalensis 7, 8, 18, 89, 90, 91, 93, 109, 113–20, 124–6, *128*, 129, 135, 144, 145, 146, 150, 151, *152*
  H. rhodesiensis 109, 110, *110*, 146
  H. rudolfensis 18, 62, 85–94
  H. sapiens 7, 7, 8, *11*, *12*, 18, *29*, 139–54, 90, *91*, 109, 123, *128*, 135, 137, 139–54
Hualongdong, China 129, 131
humans
  defining features 9–14, 83–5, 94
  evolution 8, 94, **146–7**, 152–4
  origins 83, 85–137
hunting 67, 90, 96, 114, 125, 135, 151
Huxley, Thomas Henry 94
hybridization *see* interbreeding
hyoid bones 118

ice ages 121, 123
Iho Eleru, Nigeria 149
Ilerat, Kenya 80
ilium *29*
inbreeding 126
India 144
  Narmada 110
Indonesia 122, 151
  Flores 123, 132–6
  Java 94, 96, 97, 100, 144, 145
  Sulawesi 97, 136, 145
  Sumatra 32, 144
  see also Borneo
injuries 77, *77*, 117
insular dwarfism 133, 137
interbreeding 93, 129, 144–5, 149, 150, 151, 154
Iraq
  Shanidar Cave 91
iron oxides 90
Israel *141*
  Kebara Cave 118
  Manot Cave 145
  Misliya Cave 143
  Nesher Ramla 124
  Qafzeh 91, 143
  Qesem 124
  Skhul 91, 143
  Zuttiyeh 124
Italy 114
  Ceprano 110
  Saccopastore 124
ivory 90, *91*, 92, *148*, 149

Jakovec Cavern, South Africa 38
Java, Indonesia 94, 96, 97, 100, 144, 145, 154
jaws 64, 94, 106, 109, 140
  lower *10*, *35*, *49*, 61, 64, *65*, *69*, *72*, 73, 75, 88, 99, *102*, 109, 110, 112, 129, 135, 136, 151, *151*
  upper 9, *10*, *11*, *35*, 64, 66, *67*, 75, *83*, 86, 88, 101, 131, 143
Jebel Irhoud, Morocco *139*, 140, 146
Jersey, Channel Islands 151
jewellery 91, 92, 143, 148
Jinniushan, China 129, 131
Johanson, Donald 52

Kabwe, Zambia 109, 110, *110*, 112
Kadanuumuu *60*, 60–1
Kanapoi, Kenya 49, 51
Kanjera, Kenya *17*
Kapthurin, Kenya 110
Kebara Cave, Israel 118

Kenya 52
    Allia Bay 49
    East Turkana 86, *87*, 96, 100
    Eliye Springs 140
    Guomde 141
    Ilerat 80
    Kanapoi 49, 51
    Kanjera *17*
    Kapthurin 110
    Koobi Fora 62, 80, 86
    Lake Turkana 49, 62, 86
    Lomekwi 62, 68, 83
    Nariokotome 96, *98*
    Tugen Hills 25, *25*
    West Turkana 73, *74*, 96, *98*, 140, 141, 149
Kenya National Museum
    KNM-ER 1470 86, *87*
    KNM-ER 1813 *87*
    KNM-WT 15000 *98*
Kenyanthropus platyops 19, 35, 62, *63*, *65*, 66, 69, 88
Kocabaș, Turkey 109
Koobi Fora, Kenya 62, 80, 86
Koro Toro, Chad
    KT-12/H1 61
Kow Swamp, Australia 149
Krapina, Croatia 124
Kromdraai, South Africa 75

La Cotte de St Brelade, Jersey 151
Laetoli, Tanzania 52, 57–9, *59*, 153
    Laetoli 18 cranium 140
land bridges 96, 149
language, use of 84
Laos
    Tam Pa Ling 145
Lascaux, France 93
Le Moustier, France 90
Leakey, Louis 78, 79, 85, 86
Leakey, Mary 59, 78, 85
Leakey, Meave 49, 62
Leakey, Richard 86
Ledi-Geraru, Ethiopia 88
legs *53*, 80, 85, 86, 88, 106, 125, 150
    ankle 47, 51, 54
    knee 51
    shin bone 38, *50*, 51, 54, 109, 111, *112*
    thigh bone 24, 26, *26*, 29, 44, 54, 67, 94, 101, 136, 143
    see also feet
Lehringen, Germany 90
Lesedi Chamber, South Africa 106
Levallois tool production 89, 90

Liang Bua Cave, Indonesia 132–3, *133*, 135
    Liang Bua 1 skeleton *134*
lineages
    age ranges **146–7**
    divergence 8, 9, 21, *21*, 22, 23, 113, 118, 127, 129, 146, 153
Little Foot *37*, 38–41
Lomekwi, Kenya 62, 68, 83
Lomekwian stone tools *65*, 68, *68*, 69, 83
Longlin Cave, China 149
Lucy 52, *52–3*, 60, *83*, 88
Lukeino formation 25
lumbar lordosis 44, 47
luminescence dating 33
lungs 125
Luzon, Philippines 136

Maba, China *128*, 129
Madjedbebe, Australia 145
Magdalenian 93–4
magnetic pole reversals 33, 37, 106
Makapansgat, South Africa 41, 43
Mala Balanica, Serbia 110
Malapa, South Africa 14, 37, 45, *45*, 47
    MH1 skull *46*
Malawi 79
Malaysia
    Sarawak 145
    see also Borneo
male hominins
    fossils 47, *60*, 60–1, 80, 91, 96, *98*, 113
    reconstructions *132*, *140*
Maludong Red Deer Cave, China 149
manganese 73, *74*, 90
manipulation of objects 14, 35, 47, *47*, 69
Manot Cave, Israel 145
Marine Isotope Stages 121
Mata Menge, Indonesia 136
Mauer, Germany 109, 110, 122
megadont hominins 35
Mexico *148*, 149
microcephaly 135
microliths 89, 90
Middle Awash, Ethiopia 27, 31, 66
migration
    dispersal from Africa 94–101, 143
    global spread 143–4, *148*, 149, 150–1
Milankovitch cycles 121
Misliya Cave, Israel 143

molecular clocks 8, *8*
Molefe, Nkwane 39
Morocco
    Jebel Irhoud *139*, 140, 146
Morwood, Mike 132, 135, 136
Motsumi, Stephen 39
Mousterian tool industries 90
Mrs Ples *37*
Mungo 1 145
Mungo 3 burial *144*, 145
muscles, chewing 40, 52, *72*, 75

Nariokotome, Kenya 96, *98*
Narmada, India 110
Neander Valley, Germany 113, 126
Neanderthals *see Homo neanderthalensis*
Nesher Ramla, Israel 124
New Guinea 121, 132
Ngaloba, Tanzania 140
Ngandong, Indonesia 97
Niah Cave, Malaysia 145
Nigeria
    Iho Eleru 149
nose 85, 125
Nutcracker Man 79

Oase Cave, Romania 151
Oase lower jaw 151, *151*
occipital bone *105*, 118, 141
Oldowan stone tools 68, 78, *78*, 79, 85, 86
Olduvai Gorge, Tanzania *48*, 68, 78, *78*, 81, 85
    OH5 cranium 79, *80*
    OH8 foot bones *88*
    OH62 86
    OH80 skeleton 80
Omo, Ethiopia 73
Omo Kibish 1 partial skeleton 141, *142*, 143
Omo Kibish 2 cranium 141, 143, 146
On the Origin of Species 152
orangutans 13, 30, *54*
Orrorin tugenensis 18, 21, 25–7, 153
overlapping timespans 16, 35, 41, 65–6, 133, 151, 154; *see also* coexistence

paintings, cave *see* cave art
palaeomagnetism 33, 37, 45, 106
Papua New Guinea 150
Paranthropus 35, 38, 66, 69, 71–3, 83, 86
    P. aethiopicus 19, 73, *74*

P. boisei 19, *78*, 78–80, 85
P. robustus 19, *71*, 75–7, 77
Paraustralopithecus aethiopicus 73
Paviland Cave, Wales 148
pelvis 27, *29*, 44, *53*, 54, 112, *139*; *see also* hips
Penghu, Taiwan 129
Petralona, Greece 110
Philippines 150
    Luzon 136, *137*
phytoliths 16, 48
pigments, use of 90, 91, 92, 93, 143, 145, 148
Pithecanthropus 94–5
Pontnewydd Cave, Wales 124, *124*, *125*
Portugal 113
potassium-argon dating 33
predation on hominids 14, 42–3, 51, 77, *77*, 80, 81, 105
Předmostí, Czech Republic 92
Punung, Indonesia 144

Qafzeh, Israel 91, 143
Qesem, Israel 124

radiocarbon dating 32
rafting 136
relic species 103
ribs 44, *53*, 95, 112, 125
Rift Valley, eastern Africa 23, *24*
Rising Star cave system, South Africa 101, *103*, 104, 105, 106
robust australopiths *see Paranthropus*
rock shelters *126*, 145
Romania 145, *151*
    Oase Cave 151
running 96
Russia
    Siberia 113, 123, 126, 127, *127*, *128*, 145, 150
    Sunghir 92

Saccopastore, Italy 124
sacrum *29*
Sahelanthropus tchadensis 18, *21*, 21–5, 153
Saint-Acheul, France 89
Saint-Césaire, France *152*
Sambungmacan, Indonesia 97
Sarawak, Malaysia 145
scavenging 17, 67, 99, 114
sea level changes 121
Selam 55–7, *57*

Senut, Brigitte 25
Serbia
 Mala Balanica 110
sexual dimorphism 15, 30, 60–1, 67, 76–7
Shangchen, China 96
Shanidar Cave, Iraq 91
shoulder blades 57, 60
shoulders 40–1, *52*, 55, 101, 112, 118, 135, 139
Siberia 126, 145, 150
 Altai Mountains 113, 126, 127
 Denisova Cave 123, 127, *127*, *128*
 Ust'-Ishim 150
Silberberg Grotto, South Africa 38, *39*
Sima del Elefante, Spain 106
Sima de los Huesos, Spain 104, 115, *115*, 117, *117*, 146
sites, key fossil hominin (maps) *24*, *36*, *95*, *114*; (timeline) **122–3**; *see also under named sites*
size, body *see* height
skeletons 14, *45*, 118, 149, 150
 child 47, 55–7, 92, 96, *98*
 adolescent 45, *46*, 92, 115
 adult 7, 27, *29*, *36*, 38–40, 52, *53*, *60*, 60–1, 80, 92, 113, 115, 129, 133, *134*, 141, *142*
Skhul, Israel 91, 143
skin colour 149
skulls 9, *12*, *39*, *42*, *46*, *53*, 73, *74*, 76, *100*, *102*, 106, 120, *121*, 129, 140, *141*
 black 73, *74*
 brow ridges 24, 94, 96, 112, 125, 131, 133, 139, 140, 143
 cheek bones 40, *72*, 75, 80, 86, 109
 chin 139, 141, 143
 comparisons *12*, *104–5*, *141*
 cranial vault 11, *12*, 24, 41
 cranium 22, *22*, 40, 44, 50, *51*, 62, 66, 79, 87, 96, 100, 109, 110, *110*, 131, 140, 141, 143, 145
 crests 40, 52, 66, *72*, 73, 75, 76, 76, 79, *80*
 facial projection 9, *12*, 50, 52, 62, 66, 71, 73, 75, *102*
 foramen magnum 23, *23*, 44
 forehead 44
 injury damage 77, *77*, 117
 see also jaws *and* teeth

sleeping platforms *54*, 55
Slovenia 91
social organization 30, 61, 92
social signalling 96
Solutrean phase 92
South Africa 123, 153
 Cooper's D 75
 Cradle of Humankind World Heritage Site 35, *36*, *71*
 Drimolen 69, *71*, 75, 76, 101
 Elandsfontein 110
 Florisbad 140
 Gladysvale 43
 Kromdraai 75
 Makapansgat 41, 43
 Malapa 14, 37, 45, *45*, 47
 Rising Star cave system 101, *103*, 104, 105, 106
 Sterkfontein 14, 37, *37*, 38, 43, 44, 69, 75
 Swartkrans 69, 75, 77
 Taung 41
Spain 114, 126
 Altamira Cave 93, *93*
 Atapuerca 104–5, 106, *106*, 115, *115*, 122, 123
 Gran Dolina 106, *106*, 109, 115
spears 90, *112*, 120, *149*
species, definition of 15
Steinheim, Germany 118, 124
Sterkfontein, South Africa 14, 37, *37*, 38, 43, 44, 69, 75
Stone Age 89
stone tools 89–90, 97, 106, 129, 133, 135, 136, *139*, 143, 145
 handaxes 89, 90, 111, *111*, 117, *117*, *119*, 120
 Lomekwian 65, 68, 83
 Oldowan 68, 78, *78*, 79, 85, 86
Sulawesi, Indonesia 97, 136, 145
Sumatra, Indonesia 32, 144
Sunghir, Russia 92
suprainiac fossa 118
Swanscombe, England 118, *119*, 120, *120*, 123
Swartkrans, South Africa 69, 75, 77
sympatric species 81

Taiwan
 Penghu 129
Tam Pa Ling, Laos 145
Tanzania
 Laetoli 52, 57–9, *59*, 153

Ngaloba 140
Olduvai Gorge *48*, 68, 78, *78*, 81, 85, *88*
Tasmania 121
Taung, South Africa 41
Taung child 41–3, *43*
teeth 31, *46*, 80, 94, 101, 136, 144, 145, 151
 calculus deposits 16, 48, 114
 canines 9, *10*, 11, *11*, 23, 27, 30, 41, 44, *49*, 50, 54, 56, 61, 65, 66, 71, 73, 75
 crown 27, 106
 dental arcade 11, *11*, 50, 54
 dentine 79
 and diet 16–17, *17*, 48, 51, 55, 62, 75, 79
 enamel 9, 16, 33, 42, 48, 55, 62, 71, 75, 79, 139
 gaps 9, 11, 23, 41, 44
 honing 11, 27, 31, 44, 54
 incisors 9, 11, 61, 71, 73, 75, 111, 119
 milk teeth 41, 55
 molars 11, *11*, 35, 41, 42, 62, 71, 73, 127, *128*, 129, *130*, 131, 140
 parallel rows 9, 11, *49*, 50
 premolars 11, *11*, 27, 44, 50, 61, 71, 75
 roots 65, 73, 106
 sockets 73, 86
 wear 16, *17*, 23, 44, 55, 79, 118–19
 wisdom teeth 133
Terblanche, Gert 75
'termite fishing' 68, 69, 76
Thames, river/valley *108*, 120
Tighennif, Algeria 109, 122
timelines
 hominin fossils 18–19
 key sites and climatic events **122–3**
Toba volcano, Sumatra 32
tool-making 47, **68–9**, 75, 83
tools
 bone *71*, 76, *149*
 composite 92
 importance of **89–93**
 see also stone tools
toothpicks 119
Toros-Menalla, Chad 22
Toumaï cranium 22, *22*, 23, 24
tracks *see* footprints
Transvaal Museum, South Africa 44
tree-climbing *see* climbing, adaptations for
tuff *48*, 49

Tugen Hills, Kenya 25, *25*
Turkana *see* East Turkana *and* West Turkana
Turkana Boy 96, *98*
Turkana, Lake 49, 62, 86
Turkey
 Kocabaş 109
type fossils 110, 153

United States
 Alaska 149
uranium-lead dating 45
uranium-series dating 33
Ust'-Ishim, Siberia 150

variation 13, 15, *52*, 66, 76, 86, *105*, 118, *141*, 149
vertebral column 44, 47, *53*
vocalizing abilities 118
volcanic activity 32, *48*, 49, 57, *59*, 141

Wajak Cave, Indonesia 145
Wales 113, 123
 Paviland Cave 148
 Pontnewydd Cave 124, *124*, *125*
walking
 fist 13
 knuckle 13, *13*
 upright 7, 23, *26*, 35, 40, *50*, 54, 57, 64, 67
 see also bipedalism
Wallacea 136
weight 27, *29*, 51, 61, 76, 80, 95, 99, 101, 112, 118, 125, 143, 148
West Turkana, Kenya 73, *74*, 96, *98*, 140, 141, 149
White, Tim 27, 52
Wolo Sege, Indonesia 136
Woranso-Mille, Ethiopia 50, *51*, 60, *60*, 64, 65

Xiahe, China 129, 131
Xuchang, China 129
Xujiayao, China 129

Yunxian, China 110

Zambia
 Broken Hill mine, Kabwe 109, 110, *110*, 112
Zhoukoudian Cave, China 97, *99*, 122
Zinj 79
Zinjanthropus boisei 79
Zlatý kůň, Czechia 151
Zuttiyeh, Israel 124

Picture credits

p.6, 52, 112(top), 113, 148 ©John Sibbick/Science Photo Library; p.10 ©Fiona Rogers/naturepl.com; p.12 top and bottom ©Javier Trueba/MSF/Science Photo Library, middle ©Ingo Arndt/Minden/naturepl.com; p.13 ©David Pike/naturepl.com; p.17 top ©Peter S. Ungar; bottom ©McPherron S., Alemseged Z. et al. 2010, Evidence for stone-tool-assisted consumption of animal tissues before 3.39 million years ago at Dikika, Ethiopia, Nature 466, pp.857-860; p.18, 131 (*H. daliensis* skull) ©Ji Q., Wu W., Ji Y., Li Q., and Ni X. (2021). Late Middle Pleistocene Harbin cranium represents a new Homo species. *The Innovation* 2(3), 100132; p.20 ©MPFT; p.22 ©John R. Foster/Science Photo Library; p.25 ©Ann Gibbons; p.26, 50, 57 (left), 74, 80, 88, 98, 128 (bottom), 134 (left) ©Human Origins Program, NMNH, Smithsonian Institution; p.28 ©From White T.D. et al, 2009, Ardipithecus ramidus and the Paleobiology of Early Hominids, Science 326, pp.64-86. Reprinted with permission from AAAS; p.32 ©Katerina Douka; p.30, 77, 97, 126 ©Jay Matternes; p.34, 39, 40, 58 ©John Reader/Science Photo Library; p.37 ©Goddard_Photography/istockphoto.com; p.42 ©Pascal Goetgheluck/Science Photo Library; p.43 © Cicero Moraes/CC BY-SA 3.0; p.46; Brett Eloff. Courtesy Profberger and Wits University (Own work) [GFDL (http://www.gnu.org/copyleft/fdl.html), via Wikimedia Commons; p.47 By Profberger (Own work) [CC BY-SA 3.0 (http://creativecommons.org/licenses/by-sa/3.0) or GFDL (http://www.gnu.org/copyleft/fdl.html)], via Wikimedia Commons; p.48 ©Stollhofen H. et al 2008, Fingerprinting facies of the Tuff IF marker, with implications for early hominin palaeoecology, Oldupai Gorge, Tanzania, Palaeogeography, Palaeoclimatology, Palaeoecology 259, pp. 382-409; p. 49, 63©National Museums, Kenya; p.51, 60, 64, 65 (bottom) ©Cleveland Museum of Natural History; p.54 ©Anup Shah/naturepl.com; p.56 ©Jonathan Wynn; p.57 (right) ©Photostock-Israel/Science Photo Library; p.67 ©Sailko, CC BY 3.0 <https://creativecommons.org/licenses/by/3.0>, via Wikimedia Commons; p.68 ©Harmand S. et al 2015, 3.3-million-year-old stone tools from Lomekwi 3, West Turkana, Kenya, Nature 524, pp.310-315; p.69 © de Heinzelin J. et al 1999, Environment and Behavior of 2.5-Million-Year-Old Bouri Hominids, Science 284, pp. 625-629. Reprinted with permission from AAAS; p.70 ©Louise Humphrey; p.76 ©Andy Herries, La Trobe University, Australia; p.78 By Vinny Burgoo (Own work) [CC BY 3.0 (http://creativecommons.org/licenses/by/3.0)], via Wikimedia Commons; p.82 ©Johanson, D. 2017, The paleoanthropology of Hadar, Ethiopia, Comptes Rendus Palevol 16, pp.140-154; p.89 (top) Illustration of A. afarensis by Viktor Deak © 2010 California Academy of Sciences; p.91 ©University of Tübingen/H.Jensen; p.93 © Javier trueba/MSF/Science photo library; p.99 ©Philippe Plailly/Science Photo Library; p.100 ©Georgian National Museum; p.102, 105 (right) ©Hawks J. et al 2017, New fossil remains of Homo naledi from the Lesedi Chamber, South Africa, eLife 2017;6:e24232; p.103, 104, 105 (left), 141, 152 ©Chris Stringer; p.107, 115, 116, 117 © Javier Trueba/Msf/Science Photo Library; p.111 ©Boxgrove Project; p.124, 125 ©National Museum of Wales; p.127 ©Ву Демин Алексей Барнаул (Own work) [CC BY-SA 4.0 (http://creativecommons.org/licenses/by-sa/4.0/)], via Wikimedia Commons; p.128 (top)©Dr Bence Viola; p.130 ©Dongju Zhang, CC BY-SA 4.0 <https://creativecommons.org/licenses/by-sa/4.0>, via Wikimedia Commons; p.132 ©Zhao Chuang/PNSO; p.133 By Rosino ([1]) [CC BY-SA 2.0 (http://creativecommons.org/licenses/by-sa/2.0)], via Wikimedia Commons; p.134 (right) S.Plailly/E.Daynes/Science Photo Library; p.138 ©Richter D et al 2017, The age of the hominin fossils from Jebel Irhoud, Morocco, and the origins of the Middle Stone Age, Nature 546, pp. 293-296; p.137 ©SoItchy/Shutterstock; p.142 ©Natural History Museum, London/Science Photo Library; p.151 ©Katerina Harvati

Unless otherwise stated images copyright of the Natural History Museum, London.
Every effort has been made to contact and accurately credit all copyright holders. If we have been unsuccessful, we apologise and welcome correction for future editions and reprints.

Acknowledgements

We would like to acknowledge the valuable assistance of Julia Galway-Witham, Ali Freyne and the Publishing team at the Natural History Museum.